PLASTICS LABORATORY PROCEDURES:

Identifying, Processing,
Forming, Recycling

Harry L. Hess, Ed. D.
Trenton State College
Trenton, New Jersey
And
Plastics Education
Foundation

Bobbs-Merrill Educational
Publishing Indianapolis

To my wife, Carole, and daughter,
Kimberly, for their understanding.

The Bobbs-Merrill Company, Inc.
4300 West 62nd Street
Indianapolis, Indiana 46268

While information,
recommendations, and suggestions
contained herein are accurate to the
publisher's best knowledge, no
warranty, expressed or implied, is
made with respect to the same
since the publisher exercises no
control over use or application,
which are recognized as variable.

First Edition
First Printing 1980
Cover and Interior Design by
DesignCenter, Inc.

**Library of Congress Cataloging in
Publication Data**

Hess, Harry L.
    Plastics laboratory procedures.

    1. Plastics—Laboratory
manual.   I. Title.
TP1129.H47   668.4'07'8   78-26948
ISBN 0-672-97138-0

Table of Contents

Preface

This book fills the need for a comprehensive, up-to-date, student experience-oriented guide to the many plastics education activities that may now be completed in an average school plastics laboratory. These experiences have been developed through recommendations of the Plastics Education Foundation and were thoroughly checked and edited by the Foundation before being published.

Each laboratory assignment first includes a brief review of the concepts involved in the assignment. Additionally, full and complete laboratory safety measures are outlined, with detailed reminders concerning any special safety precautions for each individual assignment. Finally, the preliminary matter in each assignment concludes with a detailed list of the equipment, tools, and materials needed to complete the assignment.

The step-by-step procedures for completing each assignment are then outlined. Each procedure is profusely illustrated. Many assignments are accompanied by a troubleshooting list to help locate and correct possible difficulties in the assignment.

The assignments conclude with a series of review questions that have been carefully written to reinforce the concepts of each exercise. A complete bibliography and material suppliers list is included with each assignment and itemizes where individual supplies may be located in the Industrial Education marketplace. This list is subject to change and addresses may be verified in most commercial/industrial directories.

Acknowledgments

The author is indebted to numerous plastics businesses and individuals who aided in the development of this book. Mentioned below are those who were very helpful.

Mr. Maurice Keroack, The Plastics Education Foundation, Albany, N.Y., for constructive criticism on the entire manuscript.

Mr. Alex Mackey Jr., Vice President, Estok Plastics Co., Trenton, N.J., for materials and assistance.

Dr. James Nichols, Assistant Professor of Industrial Arts, Trenton State College, for photographic technical assistance.

Mr. Thomas K. Rogers, Educational Consultant, Brodhead-Garrett Co., Cleveland, Ohio, for equipment and assistance.

Ms. Barbara Sutton, Industrial Arts Department, Trenton State College, for the total photographic production.

Trenton State College for allowing me the use of the plastics facilities.

Allied Chemical Co., Morristown, N.J.
Atlantic Products Corp., Trenton, N.J.
Boy Machines Inc., Plainview, L.I., N.Y.
Comet International, Inc., Elk Grove Village, Ill.
Conap, Inc., Olean, N.Y.
Contour Chemical Co., Woburn, Mass.
Dri-Print Foils Inc., Rahway, N.J.
Dynamit Nobel of America, Inc., Northvale, N.J.
Egan Machinery Co., Somerville, N.J.
General Electric Co., Laminated & Insulating Materials, Coshocton, Oh.

Gusmer Corporation, Old Bridge, N.J.

H.P. Preis Engraving Machine Co., Hillside, N.J.

Laurel Valley Mfg. Co., Lindenwold, N.J.

McNeil Akron, A Division of McNeil Corp., Akron, Oh.

Owens-Corning Fiberglas Corp., Toledo, Oh.

Princeton Chemical Research, Inc., Princeton, N.J.

Reliable Rubber & Plastic Machinery Co., Inc., North Bergen, N.J.

Smooth-On, Inc., Gillette, N.J.

Solidyne, Inc., Bay Shore, N.Y.

Standard Tool & Manufacturing Co., Kearny, N.J.

The Lannom Manufacturing Co., Tullahoma, Tn.

Uniloy-Springfield, Springfield, Mass.

Voith Fischer Plastics Machines Inc., Paramus, N.J.

Weldotron Corp., Piscataway, N.J.

Section I
Identifying and Classifying Plastics

Assignment 1
Materials Identification—Thermoplastics

OBJECTIVES

To identify different plastics as being either thermoset or thermoplastic type materials.

To identify different thermoplastics by giving them burn, specific gravity, solvent, tensile, and impact tests.

INTRODUCTION

Thermoplastics are processed by industry and by school plastics laboratories. These plastics and the products made from them must be identified. Tests reveal what plastics are used in materials. Tests also reveal the mechanical, physical, and chemical properties of plastics. This is done by testing plastics with the American Society for Testing Materials (ASTM) test procedures.

The ASTM has designed standard tests for plastics. These tests are used by industry, schools, and researchers to describe different characteristics of plastics. This information informs plastic manufacturers, molders, and customers how a plastic may perform.

Two standardized *tensile* and *impact* tests are described in this assignment. The test sample size, sample preparation, test conditions, and test procedure must be done to ASTM standards. These tests are just a few of the many ASTM tests.

Often, many characteristics of plastics can be found with less expensive equipment and nonstandard tests. These are *burn, odor, flame, solubility (solvent),* and *specific gravity* tests. Most of these tests are described in this assignment. Plastics used as test samples may be taken from discarded plastic products.

There are many tests to measure how plastics will perform. The best test of a plastic is when it is put to use.

SAFETY

The following precautions should be taken when making thermoplastic identification tests:

1. Work in a well-ventilated area. Do not breathe plastic fumes because some are toxic.

2. Work on a heat-resistant surface.

3. Wear safety glasses and heat-resistant gloves.

4. Keep a general purpose fire extinguisher in the work area.

5. Do not touch hot plastics test samples. Use pliers or tongs and heat-resistant gloves to handle them.

6. Do not drop hot plastics or plastic solvents on your skin, clothing, or the work surface.

7. Learn the safe operation of the test equipment. Keep your hands and fingers out of the test sample area of the tensile and impact test machines.

EQUIPMENT AND MATERIALS

Wear safety glasses and heat-resistant gloves when flame or electric heat is used. Also, work on a heat-resistant surface.

1. Thermoplastic and thermoset test identification.
   a. Electric soldering gun.

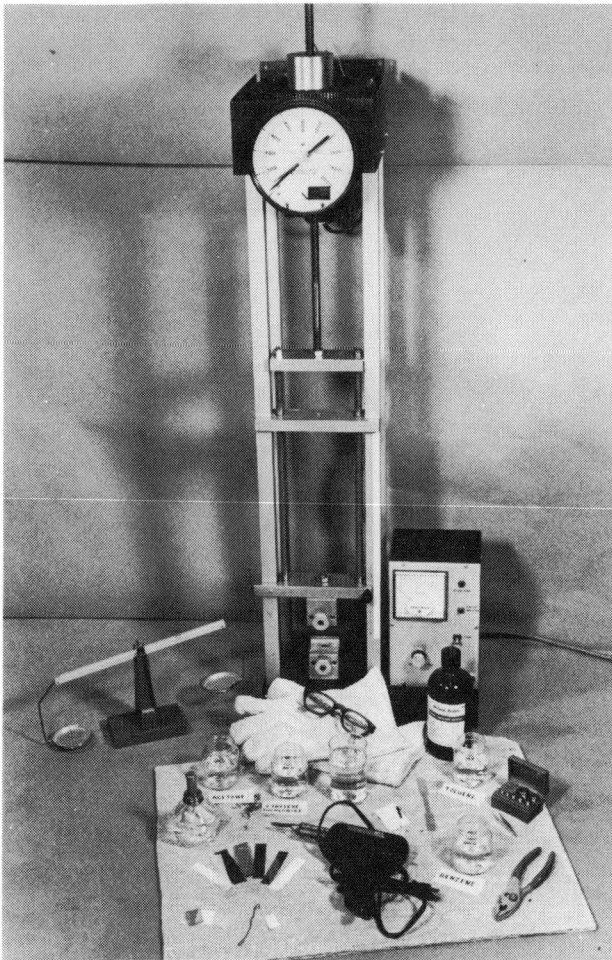

Fig. 1-1 Equipment and materials for identifying thermoplastics.

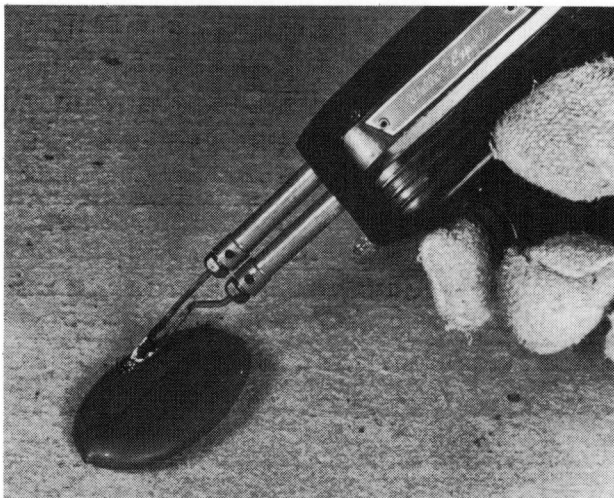

Fig. 1-2 Holding hot soldering gun against thermoplastic surface.

b. Thermoplastic product.
c. Thermoset product.
2. Plastic burn tests.
  a. Alcohol lamp or bunsen burner.
  b. Methyl alcohol fuel.
  c. Striker or matches.
  d. Pliers or tongs.
  e. Acrylic, acetal, cellulose acetate butyrate (CAB), cellulose acetate (CA), polyethylene (PE), polypropylene (PP), polystyrene (PS), polyvinyl chloride (PVC), acrylonitrile butadiene styrene (ABS), nylon, polycarbonate.
3. Solvent tests.
  a. 5 glass jars.
  b. Tweezers.
  c. Acetone, benzene, toluene, Ethylene also, dichloride.
  d. Polyethylene (PE), polystyrene (PS), nylon.
4. Specific gravity test.
  a. Scales and weights.
  b. Water container.
  c. Wire.
  d. Acrylic, polyethylene (PE), polystyrene (PS), nylon.
5. Tensile test.
  a. Tensile tester.
  b. Tensile test sample.
6. Impact test.
  a. Impact tester.
  b. Impact test sample.

BASIC LAB PROCEDURES

Obtain all the items listed under "Equipment and Materials" needed to test identify thermoplastics. See Fig. 1-1.

Thermoplastic And Thermoset Identification Test

1. Plug in the electric soldering gun. Let it heat.

2. Hold the hot soldering gun against a *thermoplastic* product for 3 or 4 seconds. See Fig. 1-2. The plastic should melt, darken, and

4

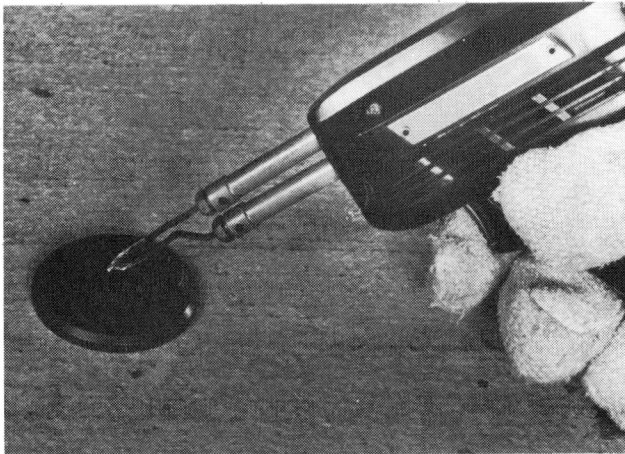

Fig. 1-3 Holding hot soldering gun against thermoset plastic surface.

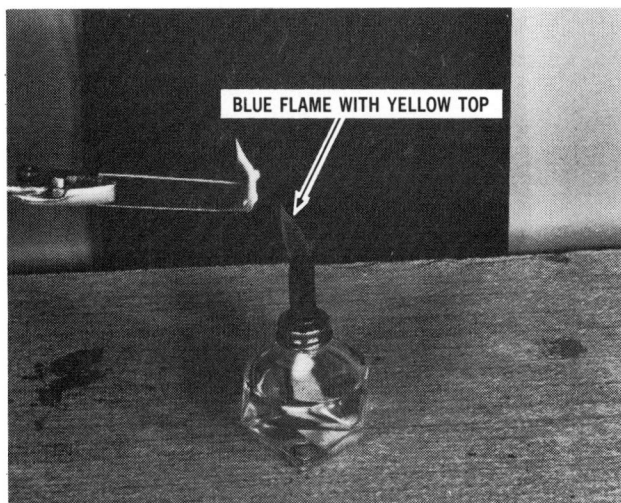

BLUE FLAME WITH YELLOW TOP

Fig. 1-4 Burn testing acrylic plastic.

BLUE FLAME WITH NO SMOKE

Fig. 1-5 Burn testing acetal plastic.

become sticky in the heated area. This test can be done to most plastics to learn if they are thermoplastics.

3. Hold the hot soldering gun against a *thermoset* product for 3 or 4 seconds. See Fig. 1-3. The plastic should char, but not melt in the heated area. This test can be done to most plastics to learn if they are thermosets.

Thermoplastic Burn Tests

1. Obtain or mold a small (⅛″ x ¾″ x 3″) acrylic sample. Follow your instructor's instructions for molding the sample.

2. Wipe the sample clean of dirt, grease, and release agent.

3. Hold one edge of the sample or part over a flame for about 10 seconds. **Be careful not to let the hot sample droppings fall into the burner.** Be sure to make all burn tests over a heat-resistant material. Keep a container of water nearby.

4. Look at the flame, smoke, and burn area. See Fig. 1-4. Acrylic is easy to burn. The burn area will soften, char, and bubble. The flame will be blue with a yellow top. Its odor will smell fruit-like. Remember that additives (chemical materials) in your plastic may cause different test results.

5. Repeat Steps 2, 3, and 4 with an acetal plastic.

6. Look at the flame, smoke, and burn area. See Fig. 1-5. Acetal is fairly easy to burn. The burn area will melt, drip, and the droppings may also burn. The flame will be blue with no smoke. Its odor will smell like formaldehyde. It will be strong and unpleasant.

7. Repeat Steps 2, 3, and 4 with cellulose acetate butyrate (CAB).

Materials Identification—Thermoplastics

Fig. 1-6 Burn testing cellulose acetate butyrate.

Fig. 1-7 Burn testing cellulose acetate.

Fig. 1-8 Burn testing polyethylene.

8. Look at the flame, smoke, and burn area. See Fig. 1-6. CAB is fairly easy to burn. The burn area will melt, drip, and the droppings may burn. The flame will be dark yellow with blue edges and black smoke. Odors will smell like rancid butter.

9. Repeat Steps 2, 3, and 4 with cellulose acetate (CA).

10. Look at the flame, smoke, and burn area. See Fig. 1-7. CA is easy to burn. The burn area will melt, drip, and the droppings may burn. The flame will be dark yellow with sooty black smoke. Odors will smell like acetic acid or burned sugar.

11. Repeat Steps 2, 3, and 4 with polyethylene (PE).

12. Look at the flame, smoke, and burn area. See Fig. 1-8. PE is easy to burn. The burn area will melt, drip, and the droppings may burn. The flame will be blue with a yellow top. Odors will smell like candle wax.

13. Repeat Steps 2, 3, and 4 with polypropylene (PP).

14. Look at the flame, smoke, and burn area. See Fig. 1-9. PP is easy to burn. The burn area will melt, swell, and drip. The flame will be blue with a yellow top and some white smoke. Odors will smell like candle wax. Note that PP and PE have about the same burn characteristics. PP burns with some white smoke and PE generally does not. Hot PE droppings may burn. Hot PP droppings generally do not burn.

15. Repeat Steps 2, 3, and 4 with polystyrene (PS).

16. Look at the flame, smoke, and burn area. See Fig. 1-10. PS is easy to burn. The burn area will soften

Fig. 1-9 Burn testing polypropylene.

Fig. 1-10 Burn testing polystyrene.

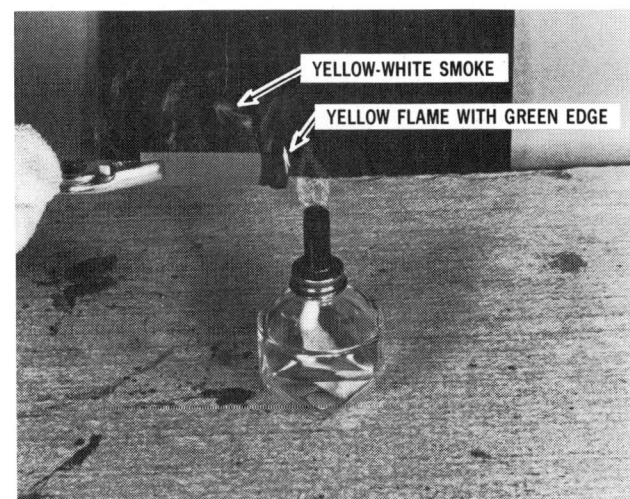

Fig. 1-11 Burn testing polyvinyl chloride.

and bubble. The flame will be orange-yellow with dense smoke and lumps of carbon. Odors will smell like acetylene gas.

17. Repeat Steps 2, 3, and 4 with polyvinyl chloride (PVC).

18. Look at the flame, smoke, and burn area. See Fig. 1-11. PVC is hard to burn and will generally put out its own flame. The burn area will soften. The flame will be yellow with a green edge, making green spurts and yellow-white smoke. *Be careful. The smoke is toxic.* Do not burn the PVC very long. Odors will smell like hydrochloric acid.

19. Repeat Steps 2, 3, and 4 with acrylonitrile butadiene styrene (ABS).

20. Look at the flame, smoke, and burn area. See Fig. 1-12. ABS is hard to burn but will burn slowly. The burn area will soften, drip, and char. The flame will be yellow with black smoke. Odors will be hard to notice.

Note that ABS is generally harder to burn than PS. The ABS burn area will often drip and PS generally does not. PS generally has more of an acetylene gas odor than does ABS. Also, PS makes carbon lumps

Fig. 1-12 Burn testing acrylonitrile butadiene styrene.

Materials Identification—Thermoplastics

Fig. 1-13 Burn testing nylon.

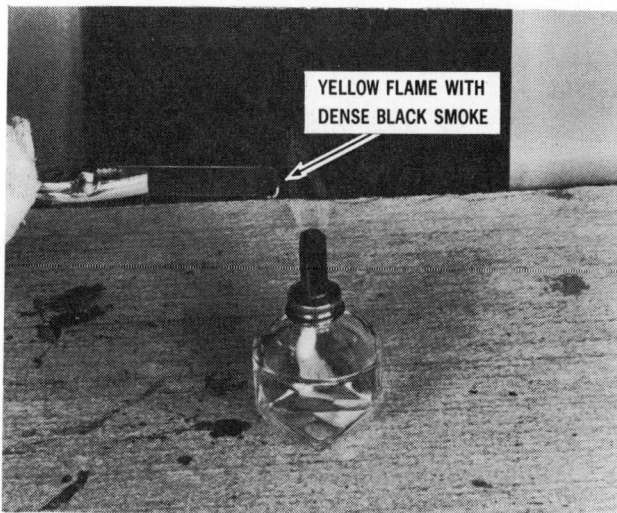

Fig. 1-14 Burn testing polycarbonate.

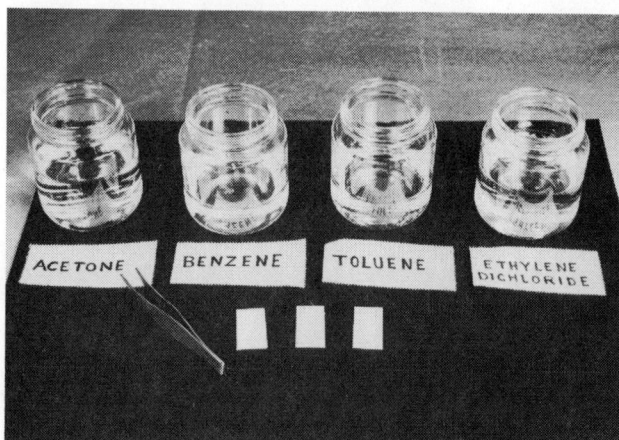

Fig. 1-15 Chemicals for testing plastics.

when it burns. ABS generally does not.

21. Repeat Steps 2, 3, and 4 with nylon.

22. Look at the flame, smoke, and burn area. See Fig. 1-13. Nylon is hard to burn and generally self-extinguishes. The burn area will burn, drip, and froth. The flame will be blue with a yellow top. Odors will smell like burned wool.

23. Repeat Steps 2, 3, and 4 with polycarbonate.

24. Look at the flame, smoke, and burn area. See Fig. 1-14. Polycarbonate is hard to burn and generally self-extinguishes. The burn area will soften, spurt, char, and break down. The flame will be yellow with dense, black smoke and carbon traces. Odors will smell like carbon.

Solvent Test

1. Place acetone, benzene, toluene, and ethylene dichloride into separate jars. See Fig. 1-15. *Handle these chemicals carefully. Some may be flammable, toxic or cause damage to your skin.*

2. Cut a small sample (⅛" x ½" x ½") of PS, PE, and nylon.

3. Dip a PS sample in the first solvent (acetone) for a few seconds. See Fig. 1-16.

4. Remove the sample to see if the plastic has started to dissolve or melt. Make sure you write down the results. Be sure to clean the tweezers with water. This should be done before placing them in another solvent. This stops one solvent from contaminating another.

5. Repeat Steps 3 and 4 with the PE sample.

Section I: Identifying, Classifying Plastics

Fig. 1-16 Dipping PS sample.

6. Repeat Steps 3 and 4 with the nylon sample.

7. Match your sample solvent results with Table 1-1. Note that PS is soluble (dissolves) in acetone, benzene, toluene, and ethylene dichloride. The PE and nylon samples will not dissolve in them. Many other solvents and plastics may be tested for solubility. This helps you to identify unknown plastics.

Plastics	Acetone	Benzene	Furfuryl Alcohol	Toluene	Special Solvents
ABS	Insoluble	Partially soluble	Insoluble	Soluble	Ethylene dichloride
Acrylic	Soluble	Soluble	Partially soluble	Soluble	Ethylene dichloride
Cellulose acetate	Soluble	Partially soluble	Soluble	Partially soluble	Acetic acid
Cellulose acetate butyrate	Soluble	Partially soluble	Soluble	Patially soluble	Ethyl acetate
Fluorocarbon	Insoluble (most)	Insoluble	Insoluble	Insoluble	Dimethyacetamide (not FEP-TFE)
Polyamide	Insoluble	Insoluble	Insoluble	Insoluble	Hot aqueous ethanol
Polycarbonate	Partially soluble	Partially soluble	Insoluble	Partially soluble	Hot benzene-toluene
Polyethylene	Insoluble	Insoluble	Insoluble	Insoluble	Hot benzene-toluene
Polypropylene	Insoluble	Insoluble	Insoluble	Insoluble	Hot benzene-toluene
Polystyrene	Soluble	Soluble	Partially soluble	Soluble	Methylene dichloride
Vinyl acetate	Soluble	Soluble		Soluble	Cyclohexanol
Vinyl chloride		Insoluble			Cyclohexanol

Table 1-1 Identification of selected plastics by solvent test method. (Courtesy Howard W. Sams Co

Fig. 1-17 Specific gravity test equipment and materials.

Specific Gravity Test

1. Set up the specific gravity test equipment. See Fig. 1-17.

2. Cut an acrylic, PE, PS, and nylon sample (⅛″ x ¾″ x 1″) to size. Make sure the sample edges are smooth. This stops air bubbles from sticking to the rough edges.

3. Drill a hole in each sample. Make the hole large enough for a fine wire to pass or hook through.

4. Weigh the wire in air (W = .63g). See Fig. 1-18.

5. Fasten the wire to the scale through a hole in one platform.

6. Weigh the wire in a container of water (X = .72g). See Fig. 1-19.

7. Fasten the wire to the sample through the drilled hole.

Materials Identification—Thermoplastics

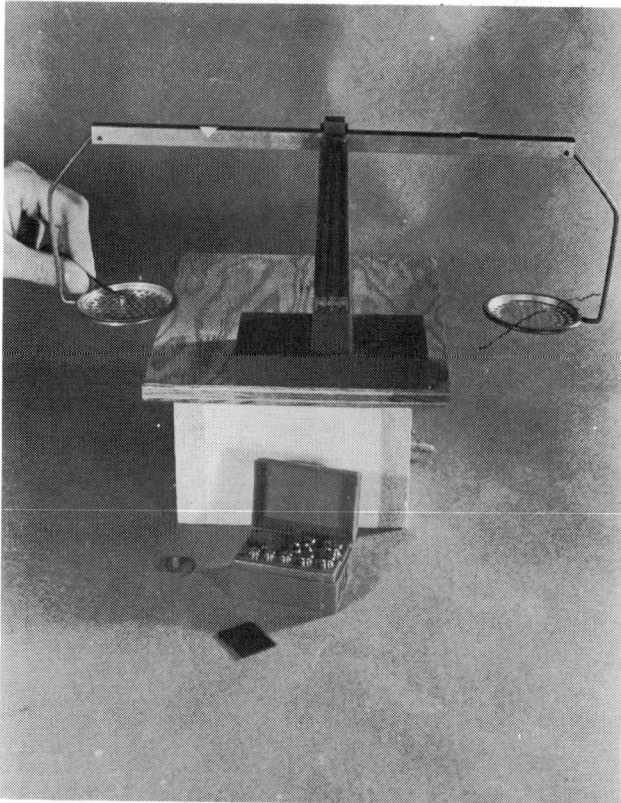

Fig. 1-18 Weighing wire in air.

Fig. 1-19 Weighing wire in water.

8. Weigh the sample and wire in air (Y = 2.72g). See Fig. 1-20.

9. Weigh the sample and wire in water (Z = .93g). See Fig. 1-21.

10. Place the above weight measurements in the formula.

$$\frac{Y - W}{(Y - W) + (X - Z)}$$

$$\frac{2.72g - .63g}{(2.72g - .63g) + (.72g - .93g)} =$$

$$\frac{2.72g - .63g}{2.09g - (.72g - .93g)} =$$

11. Solve the formula for the specific gravity of acrylic.

$$\frac{2.72g - .63g}{2.09g - (.72g - .93g)} =$$

$$\frac{2.09g}{1.88g} =$$

1.11 (specific gravity of acrylic)

The specific gravity of acrylic is 1.18; however, additives in plastics often cause off-standard test results. The weight readings may be in any weight units. Select one unit and use it for all measurements. Remember that the specific gravity of water at 73° F. is 1.0. Acrylic and nylon samples will sink because their specific gravities are greater than 1.0. The PE sample will float because its specific gravity is less than 1.0.

12. Repeat Steps 4 through 11 with a PE sample. The specific gravity of PE is about 0.91 − 0.97.

13. Repeat Steps 4 through 11 with a PS sample. The specific gravity of PS is about 0.98 − 1.1.

14. Repeat Steps 4 through 11 with a nylon sample. The specific gravity of nylon is about 1.13 − 1.20.

15. Match your specific gravity calculations with Table 1-2. Specific

Section I: Identifying, Classifying Plastics

Material	Filler or reinforce-ment	Specific gravity, g/cc	Specific weight, lb/cu in.	Specific volume, cu in./lb	Bulk factor
Phenol formaldehyde	Cellulose	1.32–1.45	0.047–0.052	21.0–19.1	2.1–4.4
	Mica	1.65–1.92	0.059–0.069	16.8–14.4	2.1–2.7
	Glass	1.60–2.20	0.058–0.079	17.3–12.6	2.0–10
	Asbestos	1.45–1.90	0.052–0.068	19.1–14.6	2.0–14
	Macerated fabric	1.36–1.43	0.048–0.051	20.4–19.4	3.5–18
Urea formaldehyde	Cellulose	1.45–1.55	0.052–0.056	19.1–17.9	2.2–30
Melamine formaldehyde	Cellulose	1.45–1.55	0.052–0.056	19.1–17.9	2.2–2.5
	Asbestos	1.70–2.00	0.061–0.072	16.3–13.9	2.1–2.5
	Glass	1.80–2.00	0.065–0.072	15.4–13.9	5.0–12
	Macerated fabric	1.50–1.55	0.054–0.056	18.5–17.9	5.0–10
Epoxy (cast, unfilled)		1.11–1.40	0.040–0.050	25.0–19.8	. . .
Molded	Glass	1.60–2.00	0.057–0.072	17.3–13.9	2.0–7
Polyester and DAP (cast, unfilled)		1.12–1.18	0.040–0.042	24.7–23.5	. . .
Molded	Clay	1.40–1.60	0.050–0.058	19.8–17.3	2.0–4
	Glass	1.35–2.30	0.049–0.083	20.5–12.0	2.0–10
Silicone	Asbestos	1.60–1.90	0.057–0.068	17.3–14.6	6.0–8
	Glass	1.68–2.00	0.060–0.072	16.5–13.9	6.0–9
Alkyd:					
Powder		1.60–2.30	0.057–0.083	17.3–12.0	1.8–2.5
Putty		1.60–2.30	0.057–0.083	17.3–12.0	1.0–1.2
ABS		1.02–1.25	0.037–0.045	27.2–22.2	1.1–1.2
Acetal		1.41–1.42	0.050–0.051	19.6–19.5	. . .
Cellulose acetate		1.20–1.34	0.044–0.048	21.9–20.7	1.8–2.6
Cellulose acetate butyrate		1.15–1.22	0.041–0.044	24.1–22.7	1.8–2.4
Cellulose nitrate		1.35–1.40	0.049–0.050	20.5–19.8	. . .
Cellulose propionate		1.17–1.24	0.042–0.045	23.7–22.4	. . .
Fluorocarbon:					
PTFE		2.10–2.22	0.075–0.080	13.2–12.5	. . .
FEP		2.12–2.17	0.076–0.078	13.0–12.8	2.0
Polyamide		1.09–1.15	0.039–0.041	25.4–24.1	1.7
Polycarbonate		1.20	0.043	23.2	1.7–5.5
Polyethylene:					
Low Density		0.910–0.925	0.032–0.033	30.4–30.0	2.0–2.5
Medium Density		0.926–0.940	0.033–0.034	29.9–29.5	. . .
High density linear		0.941–0.965	0.034–0.035	29.4–28.7	. . .
Polymethyl methacrylate		1.17–1.20	0.042–0.043	23.7–23.2	1.6–2.0
Polyphenylene oxide		1.06	0.038	26.1	2.0–3.5
Polypropylene		0.88–0.906	0.032–0.033	31.5–30.6	2.0–2.4
Polystyrene		1.04–1.100	0.037–0.040	26.6–25.2	1.6–2.4
Polysulfone		1.24	0.045	22.4	1.8–2.2
Polyvinyl acetate		1.10–1.14	0.040–0.041	25.2–24.3	. . .
Polyvinyl chloride:					
Rigid		1.35–1.45	0.049–0.052	20.5–19.1	2.0–2.3
Flexible		1.16–1.35	0.042–0.049	23.9–20.5	2.0–2.3
Polyvinylidene chloride		1.65–1.72	0.059–0.062	16.8–16.1	2.0
Polyurethane		1.11–1.25	0.040–0.045	25.0–22.2	2.4–2.8
Aluminum		2.40–2.70	0.086–0.097	11.5–10.3	. . .
Glass (plate)		2.50	0.090	11.1	. . .
Maple		0.68	0.024	40.8	. . .
Oak		0.86	0.031	32.2	. . .
Rubber (hard)		1.15–1.25	0.041–0.045	24.10–22.2	. . .
Steel		7.60–7.80	0.270–0.280	3.65–3.56	. . .

Table 1-2 Specific gravity and bulk factor of plastic materials. From *Plastics Technology by R.V. Milby.* Copyright 1973, McGraw-Hill Book Co. (Used with permission of McGraw-Hill Book Co.)

Materials Identification—Thermoplastics

Fig. 1-20 Weighing sample and wire in air.

Fig. 1-21 Weighing sample and wire in water.

gravity is the ratio of the weight of a plastic to the weight of an equal volume of water at 73° F.

Tensile Test

1. Cut or injection mold a tensile test sample to ASTM specifications. Sample sizes can be ⅛" thick × 8½" long and ¾" wide at each end. See Fig. 1-22.

2. Calculate the middle cross section (thickness x width) area of the sample.

3. Place the sample in the jaws of the tensile tester. See Fig. 1-23.

4. Move the jaws apart at 0.1, 0.2, 2.0, or 20 in. min. See Fig. 1-24.

5. Stop the jaw movement when the sample breaks.

6. Record the reading on the machine gauge head.

7. Place the gauge head reading and the cross section area in the formula.

$$\text{Tensile strength (PSI)} = \frac{\text{Maximum load (Pounds reading on gauge head)}}{\text{Original middle cross section area of sample (square inches)}}$$

8. Solve the formula and record the tensile strength.

Impact Test

1. Cut an Izod impact test sample to ASTM specifications. Sample sizes can be ⅛" x ½" x 2". See Fig. 1-25. The sample can have a 45° included angle notch 0.100" deep in its narrow edge. The notch should be cut on the ⅛" edge in the center (1" mark) of its 2" length.

2. Move the pendulum release lever to the right.

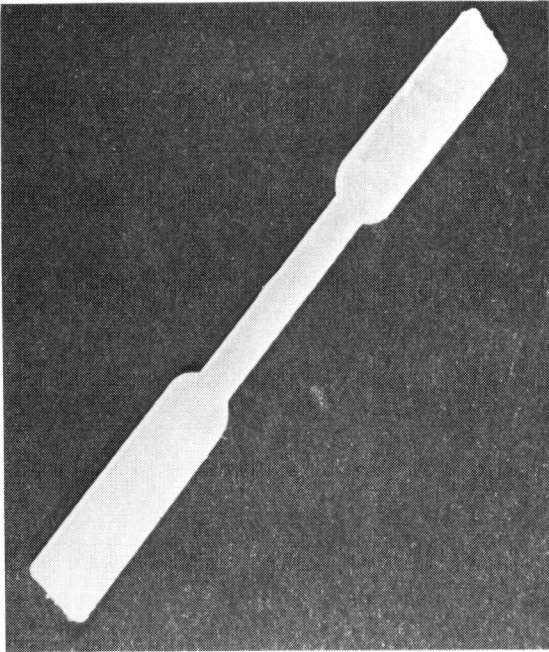

Fig. 1-22　Tensile test sample.

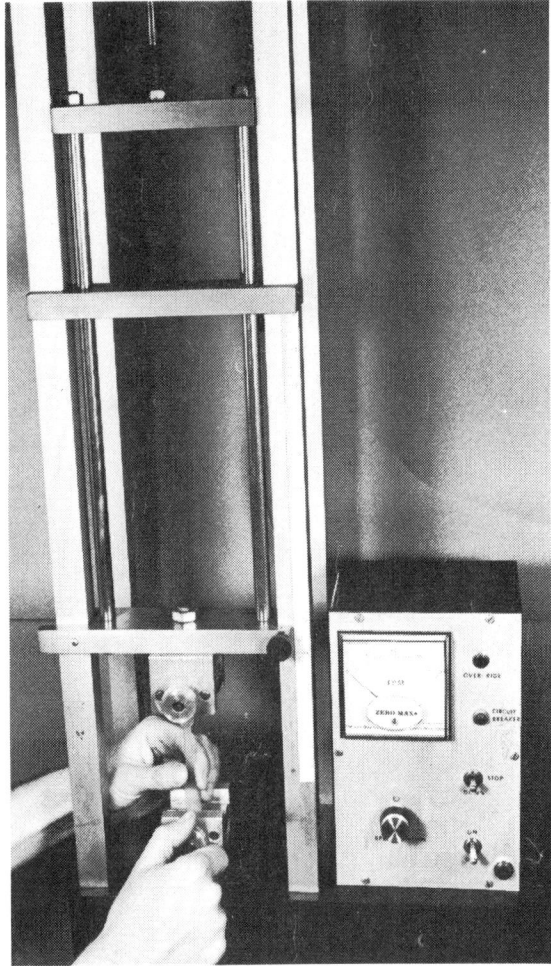

Fig. 1-23　Placing sample in tensile tester jaws.

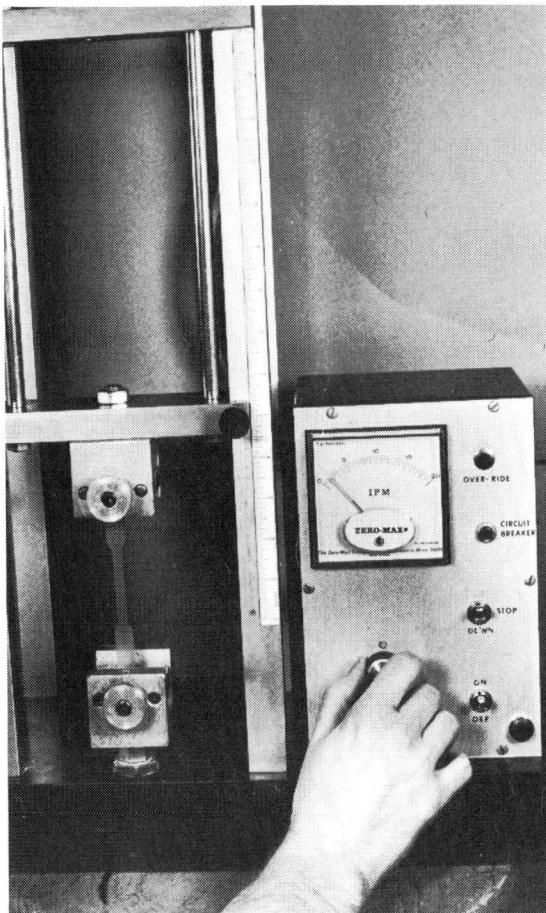

Fig. 1-24　Moving jaws apart.

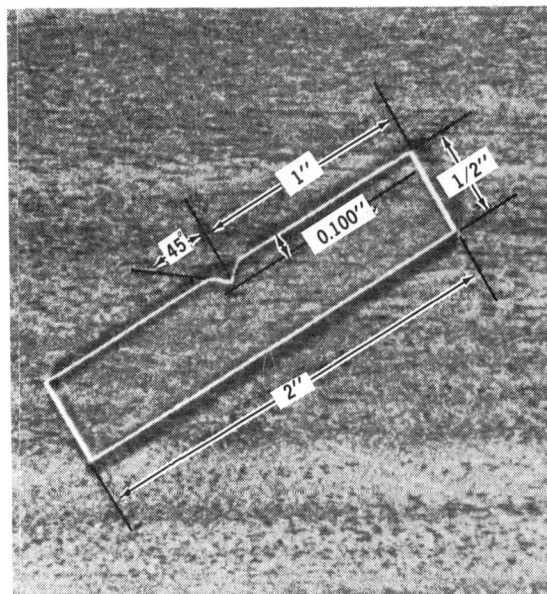

Fig. 1-25　Izod impact test sample.

Materials Identification—Thermoplastics

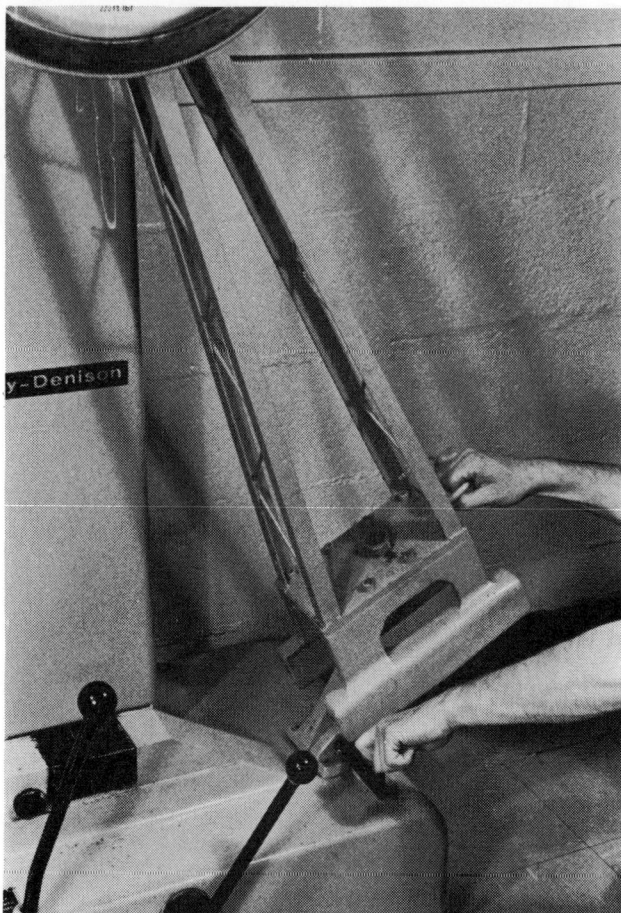

Fig. 1-26 Locking lever clear of machine jaws.

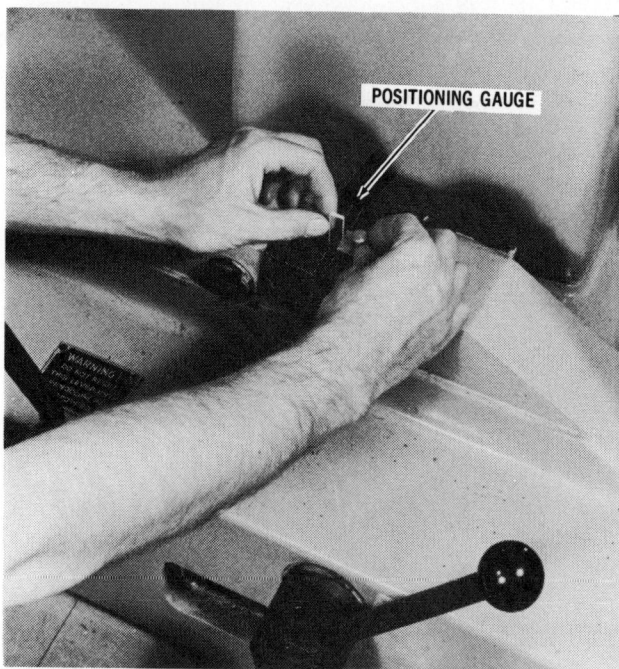

POSITIONING GAUGE

Fig. 1-27 Setting correct sample height.

3. Pull the pendulum up to the right and lock it clear of the machine jaws. See Fig. 1-26.

4. Place the sample loosely in the machine jaws with the notch facing the pendulum.

5. Set the correct sample height with the positioning gauge. Fig. 1-27. The gauge should be placed on the top of the right hand jaw. One end of the gauge should rest in the sample notch.

6. Lock the jaws. See Fig. 1-28.

7. Raise the pendulum to the proper level. It will automatically lock in place.

8. Adjust the pointer so it contacts the pendulum pointer. See Fig. 1-29.

9. Release the pendulum. See Fig. 1-30.

10. Read the foot-pounds of impact force indicated by the pointer. See Fig. 1-31. This is the Izod impact strength of the plastic. This test also can be done with a charpy plastic sample. Use the proper ASTM procedure and sample.

CONCLUSION

Specific gravity calculations do not always match the standard results. This is caused by fillers and other additives in the plastic. This makes exact thermoplastic identification hard to do. Only polyolefins, ionomers, and a few other thermoplastics have a specific gravity of less than 1.0.

Solubility (solvent) tests help to identify specific plastics. All thermoplastics do not dissolve in a specific solvent. The polyolefins, polyamides, and the fluoroplastics, are generally insoluble at room temperature. Most of the other

Section I: Identifying, Classifying Plastics

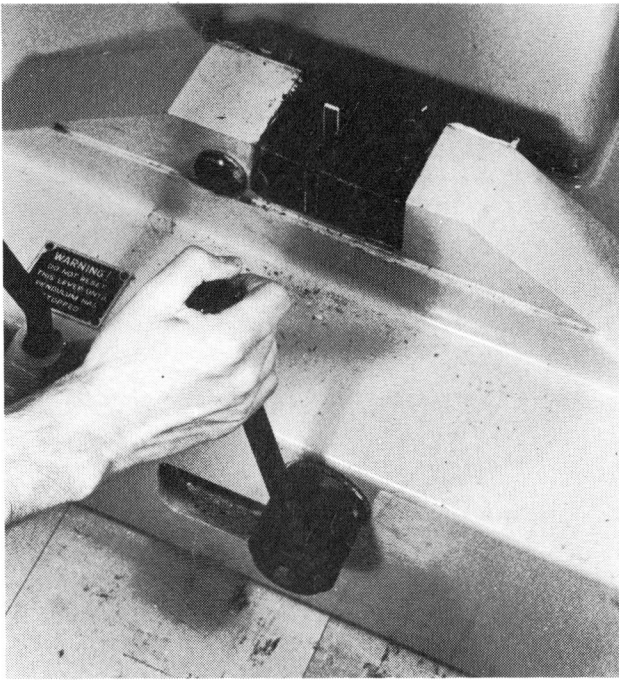

Fig. 1-28 Locking the jaws.

Fig. 1-29 Adjusting pointer.

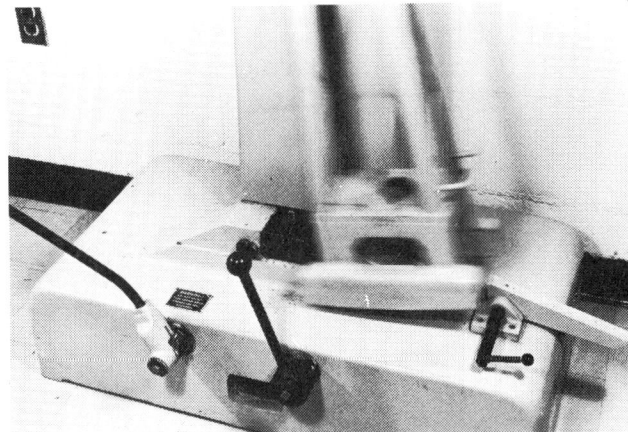

Fig. 1-30 Releasing pendulum.

thermoplastics are soluble at room temperature.

A tensile test is done by pulling a standard plastic sample apart. During the test, some plastics stretch to many times their original length. This is called "elongation." Generally, a brittle plastic will not elongate much. A more flexible plastic will. ASTM formulas can be used to calculate sample elongation. Often a tensile test machine can also be made to do flexural tests.

An impact test is made by breaking a standard plastic sample. Some thermoplastics are hard to impact test. This is because they are too soft and flexible. Some thermoplastics have an impact strength near that of steel. Examples of these are certain epoxies and polycarbonates.

Testing and identifying plastics is necessary in the school lab. The tests described in this assignment can help identify certain thermoplastics. Often the exact identification of a thermoplastic is hard to make. This is because most plastics have additives which may give wrong test results. All results given for the burn, odor, specific gravity, solubility, tensile, and impact tests were for known plastics.

Fig. 1-31 Reading indicated foot-pounds impact force.

Materials Identification—Thermoplastics

REVIEW

1. Describe each step of the specific gravity test.

2. Describe what a thermoplastic sample looks like after being heated with a hot soldering gun.

3. Describe each step for testing the solubility of PS.

4. Explain the purpose of the tensile test.

5. Describe how to burn test acrylic.

6. Describe the flame, burn area, and odor characteristics of polycarbonate.

7. Give one reason for using ASTM standardized tests.

8. Explain the purpose of the impact test.

9. Describe what additives do to burn, specific gravity, solubility, tensile, and impact test results.

10. List three plastics that will float in water.

11. Name two thermoplastics that are generally insoluble in most solvents.

12. Name one plastic that will greatly elongate during tensile testing.

13. Which thermoplastics have an impact strength close to that of steel?

SELECTED BIBLIOGRAPHY

Agranoff, Joan, ed. *Modern Plastics Encyclopedia.* New York: McGraw-Hill Book Company, 1976-77.

Baird, Ronald J. *Industrial Plastics.* South Holland, Illinois: The Goodheart-Willcox Company, Inc., 1971.

Celanese Plastics Company. *Standard Tests on Plastics.* Bulletin No. GIC, 7th Edition. Newark, New Jersey: Celanese Plastics Company, 1974.

Milby, Robert V. *Plastics Technology.* New York: McGraw-Hill Book Company, 1973.

Patton, William J. *Plastics Technology: Theory, Design, and Manufacture.* Reston, Virginia: Reston Publishing Company, Inc., 1976.

Richardson, Terry A. *Modern Industrial Plastics.* Indianapolis, Indiana: Howard W. Sams & Company, Inc., 1974.

Rosato, Dominick V., ed. *Plastics Industry Safety Handbook.* Boston, Massachusetts: Cahners Books, 1973.

Seymour, Raymond B. *Modern Plastics Technology.* Reston, Virginia: Reston Publishing Company, Inc., 1975.

EQUIPMENT AND MATERIAL SUPPLIERS

1. Ace Scientific Supply Company, Inc., 1420 East Linden Avenue, Linden, New Jersey 07036.

2. Arthur H. Thomas Company, Vine Street at Third, P.O. Box 774, Philadelphia, Pennsylvania 19105.

3. Custom Scientific Instruments, Inc., 13 Wing Drive, Whippany, New Jersey 07981.

4. Fisher Scientific Company, 1241 Ambassador Blvd., St. Louis, Missouri 63132.

5. Para Scientific Company, P.O. Box 5006, Trenton, New Jersey 08638.

6. Sargent-Welch Scientific Company, 1617 East Ball Road, Anaheim, California 92803.

7. Testing Machines, Inc., 400 Bayview Avenue, Amityville, New York 11701.

8. Tinius Olsen Testing Machine Company, Inc., Eastin Road, Willow Grove, Pennsylvania 19090.

Alcohol lamps, bunsen burners, weights, and scales can be purchased from suppliers 1, 2, 4, 5, and 6. Impact and tensile test equipment can be purchased from suppliers 3, 7, and 8.

Section I: Identifying, Classifying Plastics

Assignment 2
Materials Identification—
Thermosets

OBJECTIVES

To identify different plastics as being either thermoset or thermoplastic type materials.

To identify different thermoset plastic samples by giving them burn and specific gravity tests.

INTRODUCTION

Thermoset plastics are processed by industry and by school plastics laboratories. These materials can become mislabeled. They must be identified before making products from them. Also, the plastic in certain products must be identified. If a product's plastic is identified, specific molds can be made to process that plastic.

An accurate way to identify thermoset plastics is to use lab equipment. This is expensive. Using *burn* and *specific gravity* tests is cheaper.

Burn tests can be made with a burner, pliers, and plastic samples. The samples are burned and their flames, burn characteristics, and odors are matched to those of known samples.

Specific gravity tests can be made with direct reading specific gravity scales. They can also be made with a gradient density column. The method described in this assignment uses a weight scale, wire, and a container of water.

Many other thermoset identification tests are used by industry. The tests described in this assignment give information to **roughly** identify thermosets.

SAFETY

The following precautions should be taken when making thermoset identification tests:

1. Work in a well-ventilated area. Do not breathe plastic fumes because some are toxic.

2. Work on a heat-resistant surface.

3. Wear heat-resistant gloves and safety glasses.

4. Keep a general purpose fire extinguisher in the work area.

5. Do not touch hot plastic test samples. Use pliers or tongs and heat-resistant gloves to handle them.

6. Do not drop hot plastics on your skin, clothing, or the work surface.

EQUIPMENT AND MATERIALS

Wear safety glasses and heat-resistant gloves when flame or electric heat are used. Also, work on a heat-resistant surface.

1. Thermoset and themoplastic test identification.
   a. Electric soldering gun.
   b. Thermoset product.
   c. Thermoplastic product.
2. Plastic burn tests.
   a. Alcohol lamp or bunsen burner.
   b. Methyl alcohol fuel.
   c. Striker or matches.
   d. Pliers or tongs.
   e. Phenolic, urea, alkyd, melamine, epoxy, and polyester.
3. Specific gravity test.
   a. Scales and weights.
   b. Water container.
   c. Wire.
   d. Phenolic, melamine, and epoxy.
   e. Tweezers.

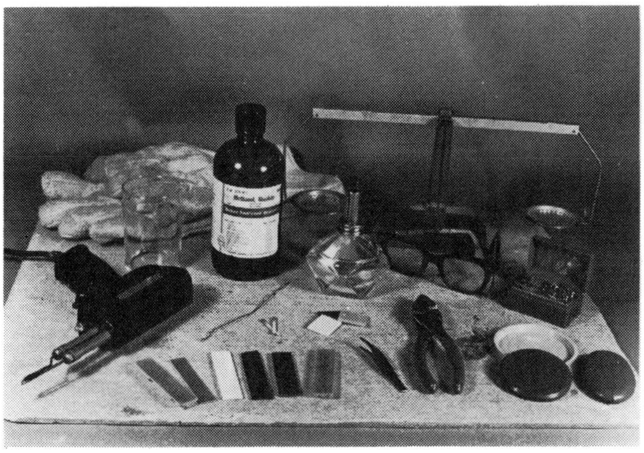

Fig. 2-1 Equipment and materials for identifying thermoset plastics.

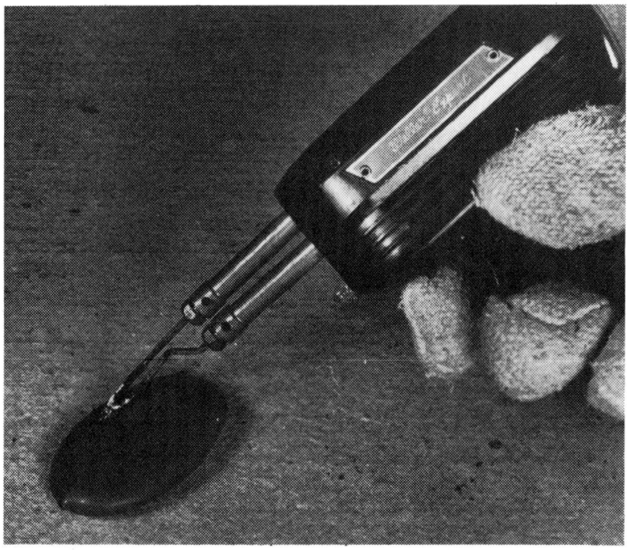

Fig. 2-2 Holding hot soldering gun against thermoplastic surface.

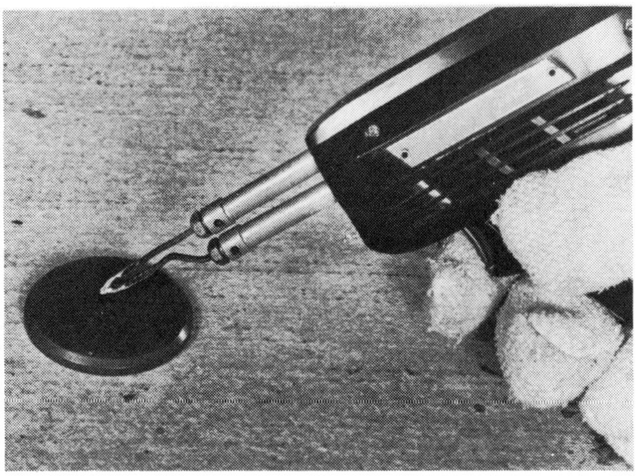

Fig. 2-3 Holding hot soldering gun against thermoset plastic surface.

BASIC LAB PROCEDURES

Obtain all the items listed under "Equipment and Materials" needed to test identify thermosets. See Fig. 2-1.

Thermoset And Thermoplastic Identificaton Test

1. Plug in the electric soldering gun. Let it heat.

2. Hold the hot soldering gun against a thermoplastic product for 3 or 4 seconds. See Fig. 2-2. The plastic should melt, darken, and become sticky in the heated area. This test can be done to most plastics to learn if they are thermoplastics.

3. Hold the hot soldering gun against a thermoset product for 3 or 4 seconds. See Fig. 2-3. The plastic should char but not melt in the heated area. This test can be done to most plastics to learn if they are thermosets.

Thermoset Burn Tests

1. Obtain or mold a small (⅛" × ¾" × 3") phenolic sample. Follow your instructor's procedure for molding the sample.

2. Wipe the sample clean of dirt, grease, and release agent.

3. Hold one edge of the sample or part over a flame for about 10 seconds. *Be careful not to let hot sample droppings fall into the burner* (bunsen or alcohol lamp). Be sure to do all burn tests on a heat-resistant surface. Keep a container of water nearby.

4. Look at the flame, smoke, and burn area. See Fig. 2-4. Phenolic is hard to burn and will self-extinguish

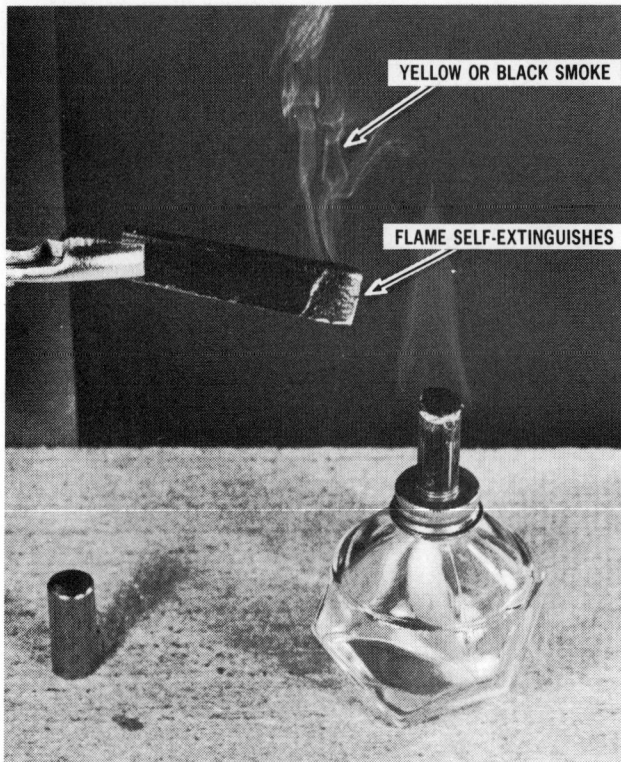

Fig. 2-4 Burn testing phenolic plastic.

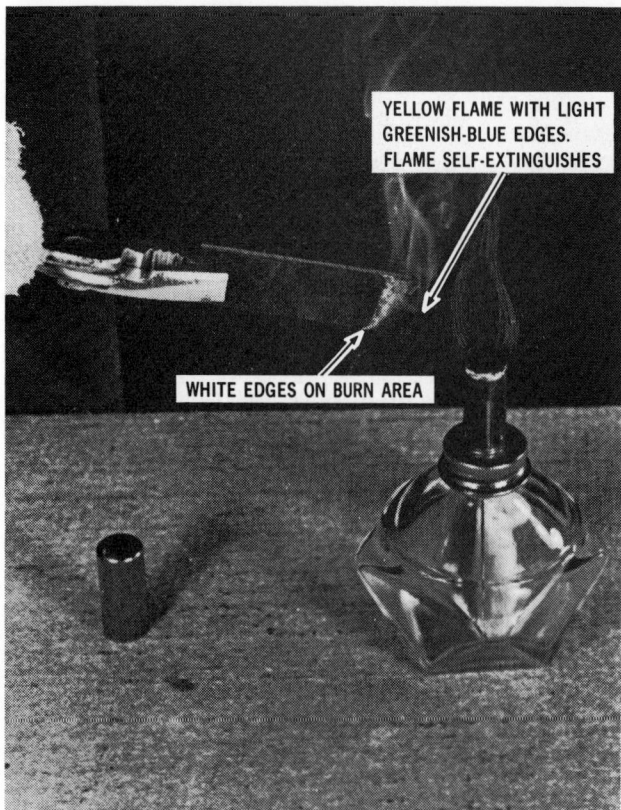

Fig. 2-5 Burn testing urea plastic.

(put its own flame out). The burn area will char, crack, swell, and often make yellow or black smoke. Odors will smell like formaldehyde or burned fabric. Remember that additives (chemicals in your plastic) may cause different test results.

5. Repeat Steps 1, 2, and 3 with a urea plastic.

6. Look at the flame, smoke, and burn area. See Fig. 2-5. Urea is hard to burn, it burns slowly and self-extinguishes. The burn area will swell, crack, and make white edges. The flame will be pale yellow with light greenish-blue edges. Odors will smell like cooked pancakes.

7. Repeat Steps 1, 2, and 3 with an alkyd plastic.

8. Look at the flame, smoke, and burn area. See Fig. 2-6. Alkyd burns steadily. The flame will be yellow with black smoke.

9. Repeat Steps 1, 2, and 3 with a melamine plastic.

10. Look at the flame, smoke, and burn area. See Fig. 2-7. Melamine is hard to burn and will self-extinguish. The burn area will swell, crack, and turn white at the edges. The flame will be light yellow. Odors will smell like formaldehyde or ammonia.

11. Repeat Steps 1, 2, and 3 with an epoxy plastic.

12. Look at the flame, smoke, and burn area. See Fig. 2-8. The burn area will char. The flame will be yellow, spurting black smoke. Odors will smell sweet.

13. Repeat Steps 1, 2, and 3 with a polyester plastic.

14. Look at the flame, smoke, and burn area. See Fig. 2-9. Polyester

Section I: Identifying, Classifying Plastics

Fig. 2-6   Burn testing alkyd plastic.

Fig. 2-7   Burn testing melamine plastic.

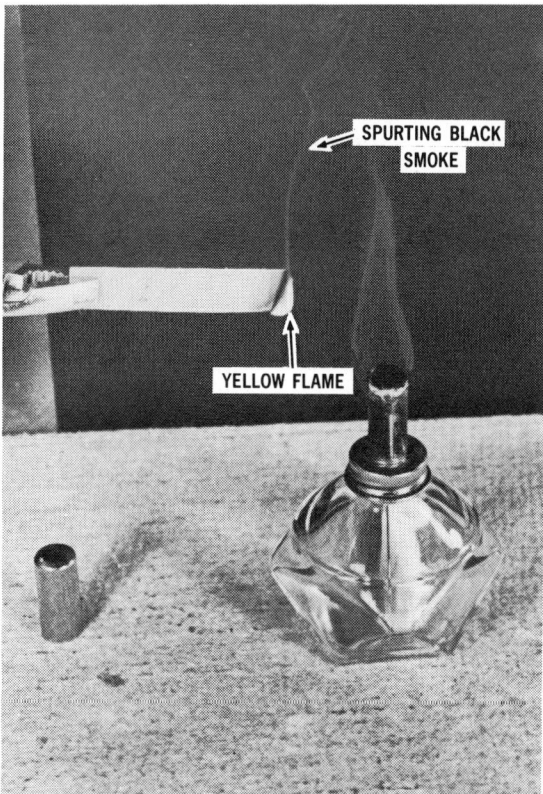

Fig. 2-8   Burn testing epoxy plastic.

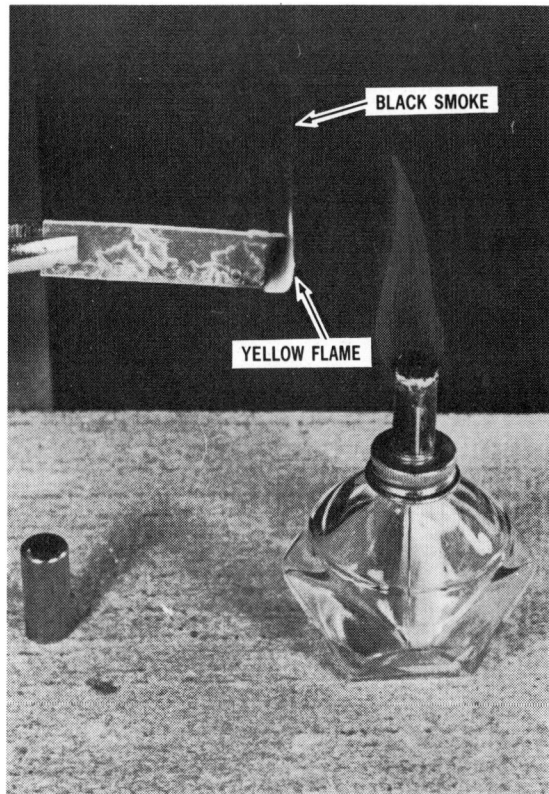

Fig. 2-9   Burn testing polyester plastic.

Materials Identification—Thermosets

will burn. The burn area softens and burns steadily. The flame will be yellow with black smoke. Odors will smell like burning coal.

Specific Gravity Test

1. Set up the specific gravity test equipment. See Fig. 2-10. A specific gravity example problem is demonstrated in Assignment 1.

2. Cut a phenolic, melamine, and epoxy sample (⅛″ × ¾″ × 1″) to size. Make sure the sample edges are smooth. This stops air bubbles from sticking to the rough edges.

3. Drill a hole in each sample. Make the hole large enough for a fine wire to pass through or hook through.

4. Weigh the wire in air (W). See Fig. 2-11.

5. Fasten the wire to the scale through a hole in one platform.

6. Weigh the wire in a container of water (X). See Fig. 2-12.

7. Fasten the wire to the sample through the drilled hole.

8. Weigh the sample and wire in air (Y). See Fig. 2-13.

9. Weigh the sample and wire in water (Z). See Fig. 2-14.

10. Place the above recorded weights in the formula.

$$\text{Specific gravity} = \frac{Y - W}{(Y - W) + (X - Z)} =$$

specific gravity of phenolic (1.25 − 1.30)

11. Solve the formula for the specific gravity of phenolic. The weight readings may be in any weight units. Select one unit and use it for all measurements. Remember that

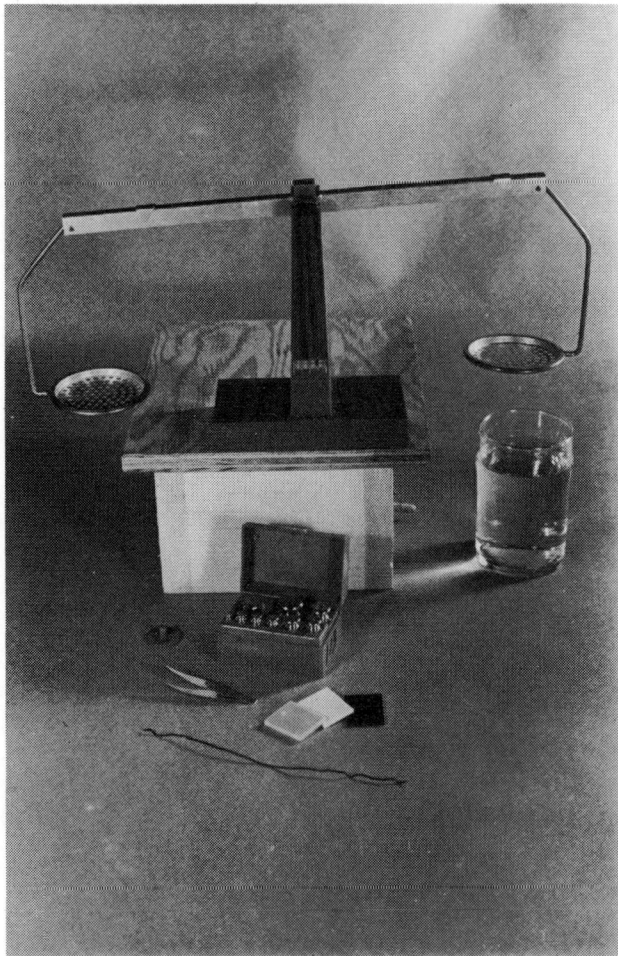

Fig. 2-10 Specific gravity test equipment and materials.

Section I: Identifying, Classifying Plastics

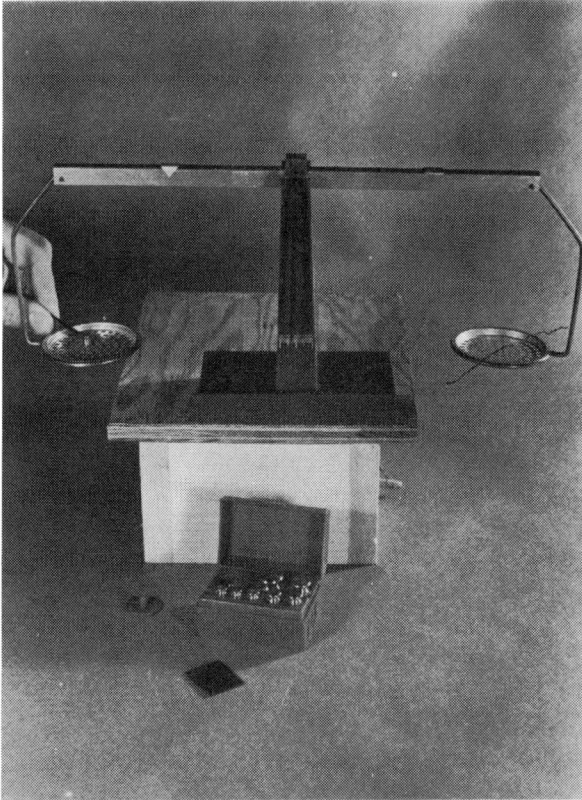

Fig. 2-11 Weighing the wire in air.

Fig. 2-12 Weighing the wire in water.

Fig. 2-13 Weighing sample and wire in air.

Fig. 2-14 Weighing sample and wire in water.

Materials Identification—Thermosets

the specific gravity of water at 73° F. is 1.0. Phenolic, melamine, and epoxy samples sink. This is because their specific gravities are greater than 1.0.

12. Repeat Steps 4 through 11 with a melamine sample. The specific gravity of melamine is about 1.48.

13. Repeat Steps 4 through 11 with an epoxy sample. The specific gravity of epoxy is about 1.11.

14. Match your specific gravity calculations with Table 2-1. Specific gravity is the ratio of the weight of a plastic compared to the weight of an equal volume of water at 73° F.

CONCLUSION

The exact identification of plastics is difficult. This is caused by the additives (chemical ingredients) in plastics. Additives often cause poor test results. Some plastics have 50% or more additives in them. When these plastics are burn tested, the additives may make the wrong color, odor, or smoke. Additives also may cause wrong specific gravity results. The burn, odor, and specific gravity tests in this assignment were made on *known* plastics.

Solubility and bend tests are not often made on thermoset plastics. Most thermoset plastics will not dissolve in many commonly used solvents.

Many thermosets are rigid and less flexible than thermoplastics. For this reason, thermosets are not often given bend and flexual tests.

When thermosets become very hot they generally char, swell, and become crumbly and brittle. Thermoplastics when heated generally become soft, sticky, and melt. This heat reaction tells us whether the plastic belongs to the thermoset or thermoplastic family.

Material	Filler or reinforcement	Specific gravity, g/cc	Specific weight, lb/cu in.	Specific volume, cu in./lb	Bulk factor
Phenol formaldehyde	Cellulose	1.32–1.45	0.047–0.052	21.0–19.1	2.1–4.4
	Mica	1.65–1.92	0.059–0.069	16.8–14.4	2.1–2.7
	Glass	1.60–2.20	0.058–0.079	17.3–12.6	2.0–10
	Asbestos	1.45–1.90	0.052–0.068	19.1–14.6	2.0–14
	Macerated fabric	1.36–1.43	0.048–0.051	20.4–19.4	3.5–18
Urea formaldehyde	Cellulose	1.45–1.55	0.052–0.056	19.1–17.9	2.2–30
Melamine formaldehyde	Cellulose	1.45–1.55	0.052–0.056	19.1–17.9	2.2–2.5
	Asbestos	1.70–2.00	0.061–0.072	16.3–13.9	2.1–2.5
	Glass	1.80–2.00	0.065–0.072	15.4–13.9	5.0–12
	Macerated fabric	1.50–1.55	0.054–0.056	18.5–17.9	5.0–10
Epoxy (cast, unfilled)		1.11–1.40	0.040–0.050	25.0–19.8	. . .
Molded	Glass	1.60–2.00	0.057–0.072	17.3–13.9	2.0-7
Polyester and DAP (cast, unfilled)		1.12–1.18	0.040–0.042	24.7–23.5	. . .
Molded	Clay	1.40–1.60	0.050–0.058	19.8–17.3	2.0–4
	Glass	1.35–2.30	0.049–0.083	20.5–12.0	2.0–10
Silicone	Asbestos	1.60–1.90	0.057–0.068	17.3–14.6	6.0–8
	Glass	1.68–2.00	0.060–0.072	16.5–13.9	6.0-9
Alkyd:					
Powder		1.60–2.30	0.057–0.083	17.3–12.0	1.8–2.5
Putty		1.60–2.30	0.057–0.083	17.3–12.0	1.0–1.2
ABS		1.02–1.25	0.037–0.045	27.2–22.2	1.1–1.2
Acetal		1.41–1.42	0.050–0.051	19.6–19.5	. . .
Cellulose acetate		1.20–1.34	0.044–0.048	21.9–20.7	1.8–2.6
Cellulose acetate butyrate		1.15–1.22	0.041–0.044	24.1–22.7	1.8–2.4
Cellulose nitrate		1.35–1.40	0.049–0.050	20.5–19.8	. . .
Cellulose propionate		1.17–1.24	0.042–0.045	23.7–22.4	. . .
Fluorocarbon:					
PTFE		2.10–2.22	0.075–0.080	13.2–12.5	. . .
FEP		2.12–2.17	0.076–0.078	13.0–12.8	2.0
Polyamide		1.09–1.15	0.039–0.041	25.4–24.1	1.7
Polycarbonate		1.20	0.043	23.2	1.7–5.5
Polyethylene:					
Low density		0.910–0.925	0.032–0.033	30.4–30.0	2.0–2.5
Medium density		0926–0.940	0.033–0.034	29.9–29.5	. . .
High density linear		0.941–0.965	0.034–0.035	29:4–28.7	. . .
Polymethyl methacrylate		1.17–1.20	0.042–0.043	23.7–23.2	1.6–2.0
Polyphenylene oxide		1.06	0.038	26.1	2.0–3.5
Polypropylene		0.88–0.906	0.032–0.033	31.5–30.6	2.0–2.4
Polystyrene		1.04–1.100	0.037–0.040	26.6–25.2	1.6–2.4
Polysulfone		1.24	0.045	22.4	1.8–2.2
Polyvinyl acetate		1.10–1.14	0.040–0.041	25.2–24.3	. . .
Polyvinyl chloride:					
Rigid		1.35–1.45	0.049–0.052	20.5–19.1	2.0–2.3
Flexible		1.16–1.35	0.042–0.049	23.9–20.5	2.0–2.3
Polyvinylidene chloride		1.65–1.72	0.059–0.062	16.8–16.1	2.0
Polyurethane		1.11–1.25	0.040–0.045	25.0–22.2	2.4–2.8
Aluminum		2.40–2.70	0.086–0.097	11.5–10.3	. . .
Glass (plate)		2.50	0.090	11.1	. . .
Maple		0.68	0.024	40.8	. . .
Oak		0.86	0.031	32.2	. . .
Rubber (hard)		1.15–1.25	0.041–0.045	24.10–22.2	. . .
Steel		7.60–7.80	0.270–0.280	3.65–3.56	. . .

Table 2-1 Specific gravity and bulk factor of plastic materials. From *Plastics Technology* by R.V. Milby. Copyright 1973, McGraw-Hill Book Co. (Used with permission of McGraw-Hill Book Co.)

Materials Identification—Thermosets

Plastics can be tested to determine if they belong to the elastomer family. Plastics belonging to this family are flexible urethane, silicones, and styrene-butadiene. To learn if a plastic is an elastomer, stretch it to twice or two and a half times its original length. If it returns to its original length when released, it is an elastomer.

REVIEW

1. Name two ways of finding the specific gravity of thermosets.

2. Describe the elastomer plastic family identification test.

3. Describe what a thermoset test sample looks like after being heated with a hot soldering gun.

4. Describe what additives do to plastic burn and specific gravity test results.

5. Explain why thermosets are not generally solvent, flexual, and bend tested.

6. Describe the specific gravity test for epoxy.

7. Define *specific gravity*.

8. Describe the procedure for burn testing a melamine sample.

9. Describe the flame, burn area, and odor characteristics for a melamine plastic.

SELECTED BIBLIOGRAPHY

Agranoff, Joan, ed. *Modern Plastics Encyclopedia.* New York: McGraw-Hill Book Company, 1976-77.

Baird, Ronald J. *Industrial Plastics.* South Holland, Illinois: The Goodheart-Willcox Company, Inc., 1971.

Celanese Plastics Company. *Standard Tests on Plastics.* Bulletin No. GIC, 7th Edition. Newark, New Jersey: Celanese Plastics Company, 1974.

Milby, Robert V. *Plastics Technology.* New York: McGraw-Hill Book Company, 1973.

Patton, William J. *Plastics Technology: Theory, Design, and Manufacture.* Reston, Virginia: Reston Publishing Company, Inc., 1976.

Richardson, Terry A. *Modern Industrial Plastics.* Indianapolis: Howard W. Sams & Company, Inc., 1974.

Rosato, Dominick V., ed. *Plastics Industry Safety Handbook.* Boston: Cahners Books, 1973.

Seymour, Raymond B. *Modern Plastics Technology.* Reston, Virginia: Reston Publishing Company, Inc., 1975.

EQUIPMENT AND MATERIAL SUPPLIERS

1. Ace Scientific Supply Company, Inc., 1420 East Linden Avenue, Linden, New Jersey 07036.

2. Arthur H. Thomas Company, Vine Street at Third, P.O. Box 774, Philadelphia, Pennsylvania 19105.

3. Fisher Scientific Company, 1241 Ambassador Blvd., St. Louis, Missouri 63132.

4. Para Scientific Company, P.O. Box 5006, Trenton, New Jersey 08638.

5. Sargent-Welch Scientific Company, 1617 East Ball Road, Anaheim, California 92803.

Bunsen burners, alcohol lamps, scales, and weights can be purchased from any of the suppliers listed above.

Materials Identification—Thermosets

Section II
Molding

Assignment 3
Injection Molding

OBJECTIVES

To prepare an injection mold.

To adjust an injection molding machine.

To make an injection molded product.

INTRODUCTION

Injection molding is an industrial process. Products are made by softening plastic and forcing it under pressure into a closed mold. The plastic cools in the mold and forms a solid product. The injection cycle is very fast. It is able to mass produce detailed products at low costs.

To begin the injection process, plastic pellets are placed in a hopper. They then feed into a heated barrel at the rear of the injection machine. The pellets are fed forward by a plunger or a screw. As the plastic moves forward in the heated barrel, it softens. When the softened plastic reaches the nozzle, the plunger forces it into a mold cavity. (A reciprocating screw is used in some machines to force the plastic into the mold.) Once in the mold, the plastic cools and the part is formed. The mold opens and ejects the part. The cycle repeats.

A small, school laboratory injection molder is shown in Fig. 3-1. This machine is the vertical plunger type. Illustrated in Fig. 3-2 is a small to medium size industrial injection machine. It is a reciprocating screw type machine.

Injection molding is used by industry for the mass production of both small (tiny gears) and large (garbage containers) parts. It is the

Fig. 3-1   School laboratory injection molder.

Fig. 3-2   Industrial injection molding machine.
(Courtesy Boy Machines, Inc.)

main molding process used with thermoplastics. A few thermosets are also injection molded.

Fig. 3-3 illustrates many injection molds. Industrial injection molds are made from carbon and alloy steels. These materials can take the high pressure produced during injection molding. Laboratory molds are made from aluminum, steel, or epoxy steel composits.

SAFETY

The following precautions should be taken when injection molding:

1. Work in a well-ventilated area. Do not breathe plastic fumes because some are toxic.

2. Work on a heat-resistant surface.

3. Wear safety glasses and heat-resistant gloves.

4. Keep a general purpose fire extinguisher in the work area.

5. Learn the safe operation of the injection molding machine.

6. Make sure the safety gate and all of the safety switches are working before you use the injection molding machine.

7. Make sure all machine control and moving part guards are in place and working.

8. Do not burn yourself on the hot injection machine nozzle and barrel.

9. Do not put your hands and arms into the injection machine hopper.

10. Be careful not to pinch your fingers or arms between the clamp jaws or mold halves.

11. Do not touch hot plastic.

Fig. 3-3 Examples of injection molds.

Injection Molding

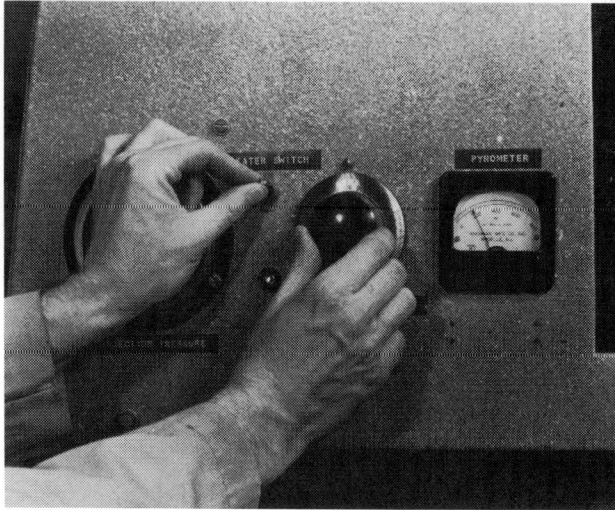

Fig. 3-4 Adjusting heat regulator dial.

Fig. 3-5 Filling hopper with plastic.

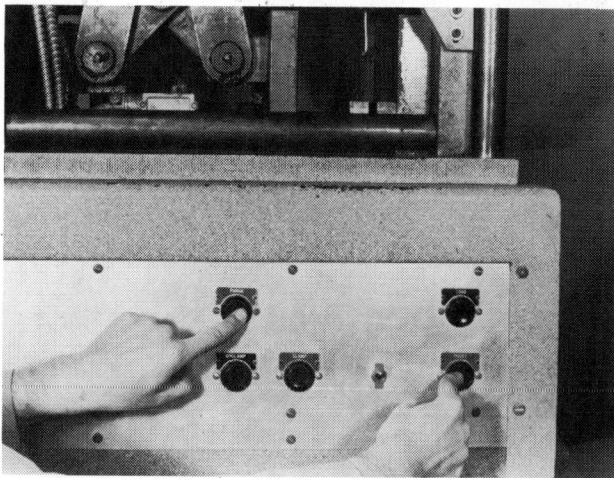

Fig. 3-6 Pushing purge and inject switches.

EQUIPMENT AND MATERIALS

The equipment and materials needed for injection molding are:

1. Safety glasses.
2. Heat-resistant gloves.
3. Heat-resistant surface.
4. Injection machine.
5. Molds.
6. Polyethylene, polypropylene, or polystyrene.
7. Mold release agent.
8. Cloth or paper towels.
9. Scoop.
10. Brass rod.
11. Utility knife and trim board.

INJECTION MOLDING LAB PROCEDURE

1. Obtain all the items listed under "Equipment and Materials" needed for injection molding.

2. Turn on the injection machine main water cooling valve. Some machines do not have this device. Make sure water is flowing through all cooling lines.

3. Switch on the injection machine heaters.

4. Adjust the heat regulator dial. See Fig. 3-4. This adjustment will be different for each type of plastic injected. Your instructor will give you the proper setting. A setting of 100 was used for this lab assignment. This setting gives a 390° F. operating temperature for the polystyrene.

5. Let the machine heat at this temperature for 15 to 20 minutes. This gives the inside machine parts (plunger and barrel) time to completely heat. Do not turn on the injection machine before the warm up period is over. If the plunger is

Fig. 3-7 Plastic forced from nozzle (purging).

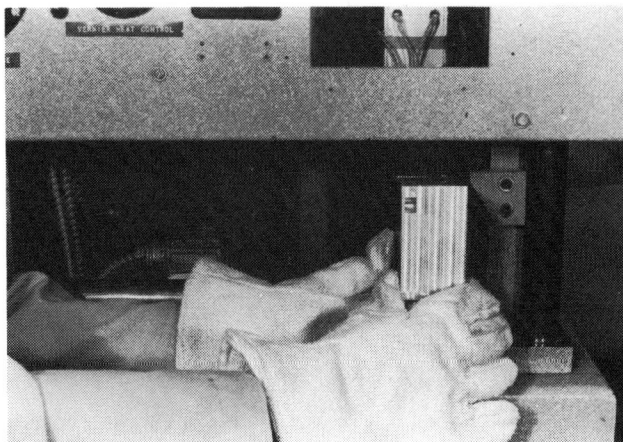

Fig. 3-8 Placing mold in machine.

moved too soon, it may break. It would also scar the barrel.

6. Check the machine gauge to make sure the correct operating temperature was reached. **Do not do the next step until the correct temperature is reached.**

7. Fill the hopper with plastic. Do this only after the machine temperature has stabilized. See Fig. 3-5.

8. Turn on the hydraulic pump motor. Some machines do not have this device.

9. Place the purge bar in position on the machine. Some machines do not have a purge bar.

10. Push the purge and inject switches. See Fig. 3-6. This will make the plunger move and force plastic out of the nozzle. See Fig. 3-7.

11. Push the reload switch until the plunger returns to its original position.

12. Repeat steps 10 and 11 until all the dirt or old plastic is out of the plastic shot stream.

13. Remove the purge bar from the machine. This will allow the nozzle to touch the mold during the injection cycle.

14. Wipe the mold cavity clean. This is the body mold for a name tag.

15. Place a mold release on the inside mold cavity.

16. Assemble the mold halves.

17. Place the mold in the machine. See Fig. 3-8. **Make sure the mold sprue lines up directly under the nozzle.**

18. Push the clamp switch until the

Injection Molding

Fig. 3-9   Clamping the mold.

Fig. 3-10   Pushing the inject switch.

mold is tightly clamped. See Fig. 3-9.

19. Push the inject switch. See Fig. 3-10. This will make the plunger move and force plastic into the mold.

20. Let the plastic cool in the closed mold. Ask your instructor how long to cool the plastic.

21. Push the mold open switch. Let the clamp completely open.

22. Turn on the safety switch. See Fig. 3-11. This will lock the clamp in the open position.

23. Remove the mold from the clamp opening. See Fig. 3-12.

24. Pull the injected name tag from the mold cavity. See Fig. 3-13. If the product sticks in the mold, use a mold release agent, place more draft in the mold, or rework the ejection device. *Do not burn your fingers on the hot mold and plastic.* Wear heat-resistant gloves when handling the mold or product.

25. Check the size and density of the product. If it is oversize, inject at a lower pressure the next time. If it is undersize, inject at a lower temperature. Inject at higher pressures or tighten the clamp if the product is not dense enough.

26. Check the product for completeness. If the product is not complete, increase the injection and clamp pressure, use more charge, increase the injection temperature, line up the nozzle with the sprue bushing or opening, or vent the mold.

27. Check the product for dirt. If the product is dirty, purge or clean the barrel, screw, or plunger of the injection machine.

28. Twist the runner system from the molded name tag. See Fig. 3-14.

Section II: Molding

Fig. 3-11 Turning on safety switch.

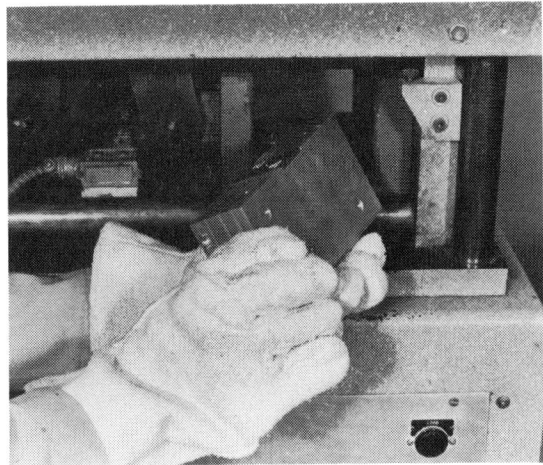

Fig. 3-12 Removing mold.

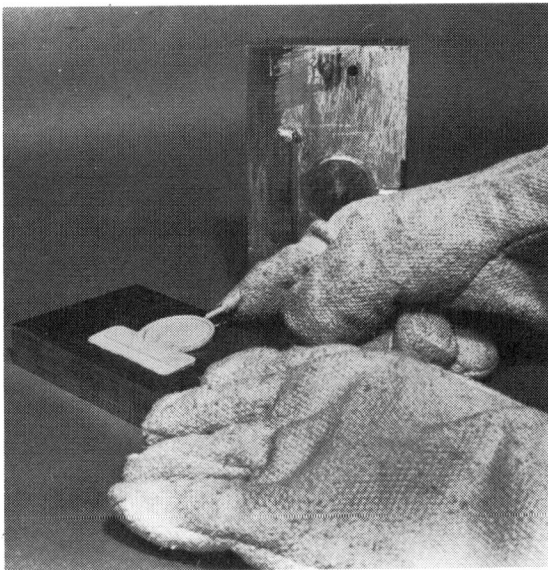

Fig. 3-13 Pulling injected name tag from mold cavity.

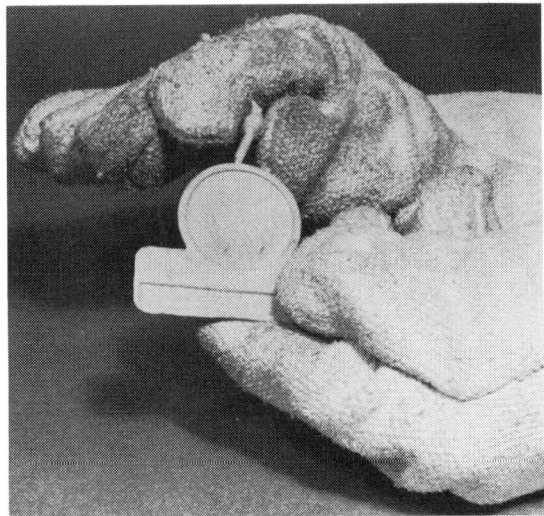

Fig. 3-14 Twisting off runner.

Injection Molding

29. Trim the flash from the product. See Fig. 3-15. Use a sharp utility knife. *Be careful not to cut yourself.* (Excess flash is caused by too little clamp pressure or too much injection pressure.) The name tag is now ready to receive a pinback, injected and hot stamped circular insert, and tape written name.

CONCLUSION

Fig. 3-16 shows many injection molded products. Use a different mold and the procedure just described to make these items.

When thermoplastics are injection molded, all of the clean, defective parts can be reused. The defective parts are collected and ground into pellets with special grinding machines. These pellets are mixed with new thermoplastics of the same type. They are recharged back into the hopper.

Injection machine sizes are calculated in two ways. One way is by using the mold clamping pressure. This is generally measured in tons. A second way is by the weight or volume of the largest plastic shot the machine can inject.

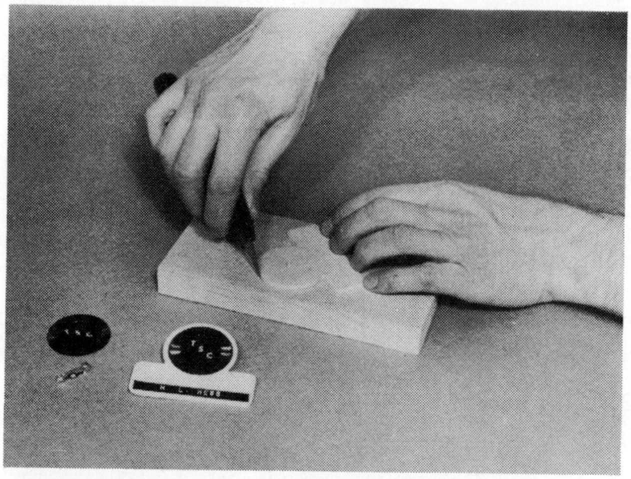

REVIEW

1. Define *injection molding*.

2. Explain each step in the injection molding cycle.

3. Name the type of plastic used in this activity.

4. Name two materials used to make injection molds.

5. Describe the process for reusing thermoplastic scrap made while injection molding.

6. Describe two ways of

Fig. 3-15 Trimming flash and gate.

Fig. 3-16 Examples of injection molded products.

Section II: Molding

determining the size of injection molding machines.

7. Define *purging*.

8. List four products made by the injection molding process.

SELECTED BIBLIOGRAPHY

Agranoff, Joan, ed. *Modern Plastics Encyclopedia.* New York: McGraw-Hill Book Company, 1976-77.

Baird, Ronald J. *Industrial Plastics.* South Holland, Illinois: The Goodheart-Willcox Company, Inc., 1971.

Injection Molding—Teacher's Manual. Indianapolis: Howard W. Sams & Company, Inc., 1974.

Milby, Robert V. *Plastics Technology.* New York: McGraw-Hill Book Company, 1975.

Patton, William J. *Plastics Technology: Theory, Design, and Manufacture.* Reston, Virginia: Reston Publishing Company, Inc., 1976.

Richardson, Terry A. *Modern Industrial Plastics.* Indianapolis: Howard W. Sams & Company, Inc., 1974.

Rosato, Dominick V., ed. *Plastics Industry Safety Handbook.* Boston: Cahners Books, 1973.

Weir, Clifford L. *Introduction to Injection Molding.* Greenwich, Connecticut: Society of Plastics Engineers, Inc., 1975.

EQUIPMENT AND MATERIAL SUPPLIERS

1. Brodhead-Garrett, 4560 East 71st Street, Cleveland, Ohio 44105.

2. Brown Plastics Engineering Company, Inc., 1823 Holste Road, Northbrook, Illinois 60062.

3. Cope Plastics, Inc., 4441 Industrial Drive, Godfrey, Illinois 62035.

4. Delvie's Plastics, Inc., 2320 South West Temple, P.O. Box 1415, Salt Lake City, Utah 84110.

5. Graves-Humphreys, Inc., 1948 Franklin Road, P.O. Box 1347, Roanoke, Virginia 24033.

6. Industrial Arts Supply Company, 5724 W. 36th St., Minneapolis, Minnesota 55408.

7. McKilligan Industrial Supply Corporation, 494 Chenango Street, Binghamton, New York 13901.

8. Paxton/Patterson, 5719 West 65th Street, Chicago, Illinois 60638.

9. Pitsco, P.O. Box 26, Pittsburg, Kansas 66762.

Injection machines are available from all of the suppliers listed above. Plastic and molds can be purchased from suppliers 1 and 3-9.

Fig. 4-1   School laboratory rotational molding machine.

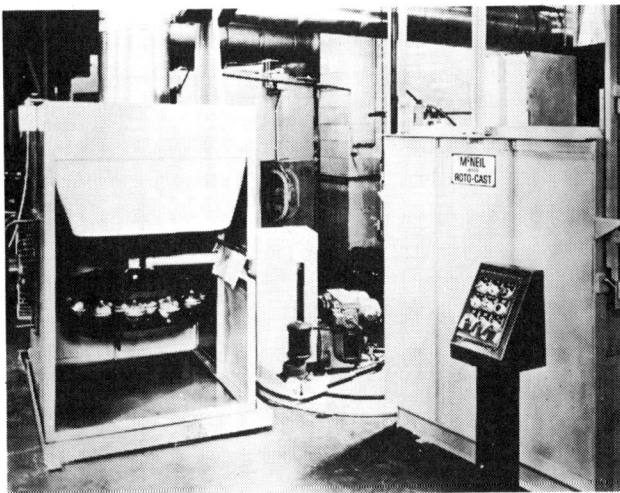

Fig. 4-2   Medium-size industrial rotational molding machine. (Courtesy McNeil Akron Division/McNeil Corporation)

Assignment 4
Rotational Molding

OBJECTIVES

To select, clean, and charge a rotational mold.

To load and adjust a rotational molding machine.

To produce a rotationally molded product.

INTRODUCTION

Rotational molding is an industrial process. Plastic powders or liquids are heated in molds. They are rotated at the same time in two planes at right angles to each other. Rigid or soft hollow plastic products are made by this process.

During molding, plastic particles melt on the inner surfaces of the hot molds. This action makes the plastic weld into a product. The wall thickness of the product is controlled by the amount of plastic placed in the mold and by the speed each mold is rotated.

Shown in Fig. 4-1 is a small school laboratory rotational molding machine. Fig. 4-2 shows a medium-size industrial rotational molding setup. In Fig. 4-3, a large-size industrial rotational molding setup is shown.

Fuel tanks, open containers, drums, and other large products are rotationally molded. Small products, such as toys, floats, car arm rests, and world globes, can also be made by rotational molding.

SAFETY

The following precautions should be taken when rotational molding:

Fig. 4-3 Large-size industrial rotational molding machine. (Courtesy McNeil Akron Division/McNeil Corporation)

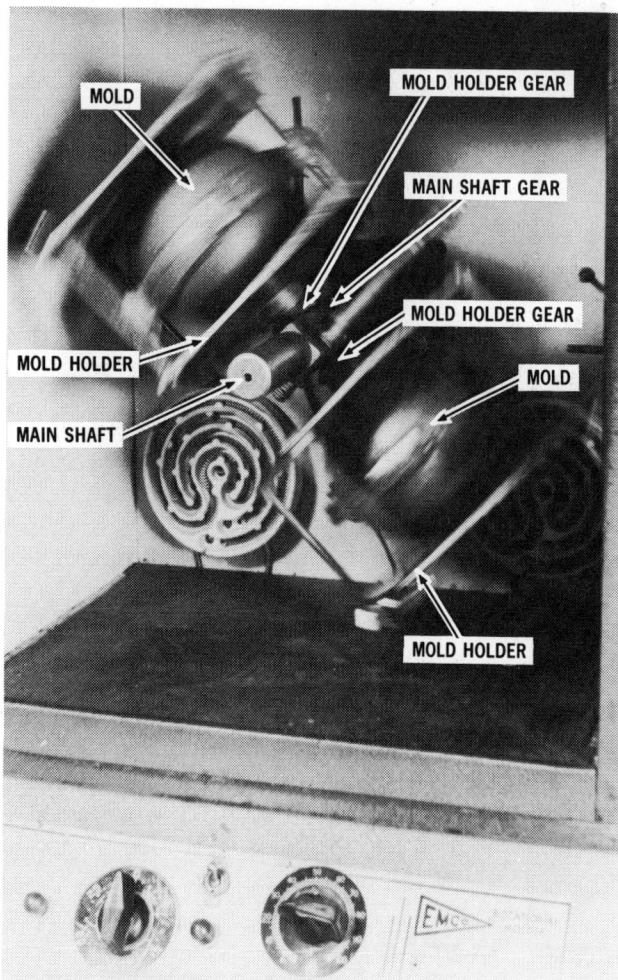

Fig. 4-4 Basic components of small rotational molding machine.

1. Work in a well-ventilated area. Do not breathe plastic fumes because some are toxic.

2. Work on a heat-resistant surface.

3. Wear safety glasses and heat-resistant gloves.

4. Keep a general purpose fire extinguisher in the work area.

5. Do not touch hot plastics, molds, or machine parts with your bare hands.

6. Learn the safe operation of the rotational molding machine.

7. Keep your hands and fingers away from rotating machine parts.

EQUIPMENT AND MATERIALS

The equipment and materials needed for rotational molding are:

1. Safety glasses.
2. Heat-resistant gloves.
3. Heat-resistant surface.
4. Rotational molding machine.
5. Molds.
6. PVC, polyethylene, ethylene vinyl acetate, or polypropylene.
7. Mold release agent.
8. Cloth or paper towels.
9. Pliers.
10. Utility knife.
11. Screwdriver.
12. 16 oz. container.

ROTATIONAL MOLDING LAB PROCEDURE

Shown in Fig. 4-4 is the inside of a small rotational molding machine. The main shaft gear meshes with a gear on each mold holder (mold holder gear). When the main shaft rotates, its gear causes the two holder gears to rotate. This gear action causes the two molds to

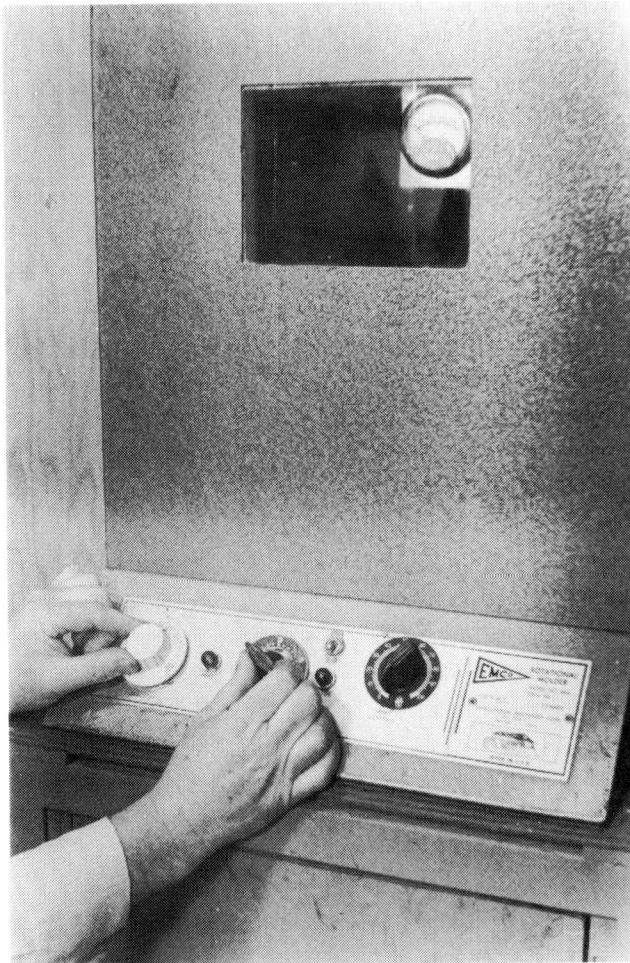

Fig. 4-5 Setting heater and timer switches.

Fig. 4-6 Charging mold.

revolve around the main shaft. The molds rotate two directions at once.

1. Set the heater and timer switches. See Fig. 4-5. The instructor will tell you the heat and timer setting. A full timer setting and a 375°F. heater was used to mold the polyethylene balls.

2. Wipe clean each rotational mold cavity (ball molds). Make sure each mold half parting surface is clean. Be sure the halves fit tightly.

3. Apply a release agent to each mold half.

4. Weigh the plastic for each mold. Your instructor will tell you how much plastic to weigh.

5. Charge one part of each two part mold with plastic. See Fig. 4-6. Powdered polyethylene will make a rigid or semi-rigid product. A plastisol (vinyl dispersion) charge will make a flexible product.

6. Clamp the mold halves tightly together with nuts and bolts, as in Fig. 4-7. **Be sure the lines or marks at the parting surface of each mold half are lined up.** If the mold marks do not line up, the product will be out of round.

7. Place each charged mold in the center of each mold holder.

8. Tighten each mold holder wing nut. See Fig. 4-8.

9. Slide the mold holder shaft over the main machine shaft, as in Fig. 4-9.

10. Turn the locking handle until it is in the locked position. See Fig. 4-10. This fastens the mold holders in the machine.

11. Close and lock the rotational machine door.

Rotational Molding

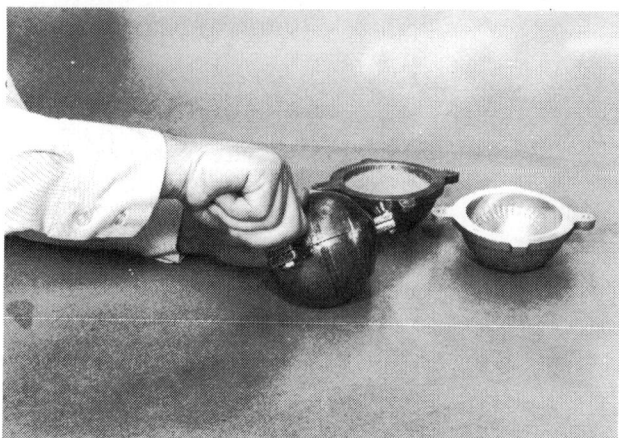

Fig. 4-7 Clamping mold halves together.

Fig. 4-8 Tightening mold holder wing nut.

12. Set the main mold axis speed control. Use the setting given by your instructor. A mid-range RPM or speed was used to mold the polyethylene balls.

13. Turn on the rotational machine switch. This starts the molds rotating. The molds rotate around two axes at one time. Each axis is at right angles to the other. This rotation causes the plastic to be spread against the total mold surface.

14. Allow the molds to rotate at the previously set molding temperatures for 4 to 5 minutes.

15. Turn off the heater and timer.

16. Open the molding machine door. Wear heat-resistant gloves. *Be careful not to burn yourself on the hot door and the gases from the open door.* Allow the molds to rotate with the oven door open for 5 minutes. This is the beginning of the mold cooling cycle.

17. Turn off the rotational mold machine motor.

18. Remove the mold holder from the main machine shaft. See Fig. 4-11.

19. Loosen the mold holder clamps. See Fig. 4-12.

20. Remove each mold from the holder. *Be careful not to burn your hands.*

21. Cool the molds in cold water for 4 to 5 minutes.

22. Unclamp each mold half.

23. Pry each mold open with a screwdriver inserted between the mold pry tabs. See Fig. 4-13. **Do not touch the mold parting or inside surfaces with metal tools.** This will scratch the molds.

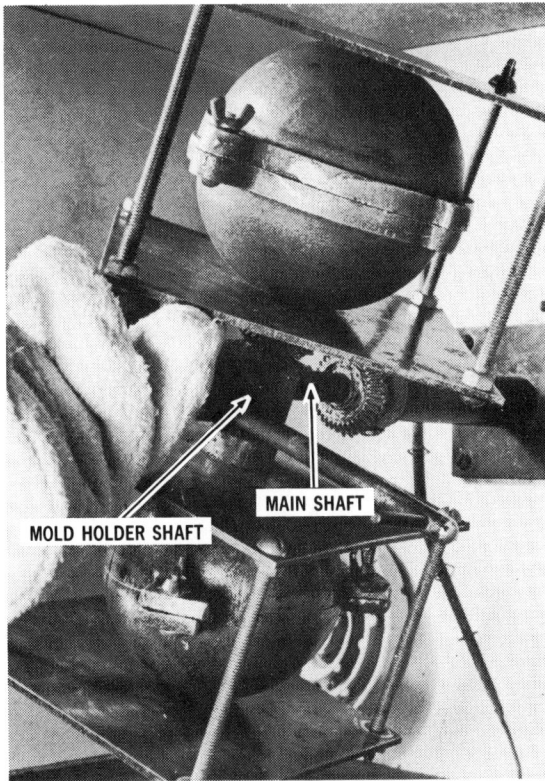

Fig. 4-9 Sliding mold holder shaft over main machine shaft.

MOLD HOLDER SHAFT

MAIN SHAFT

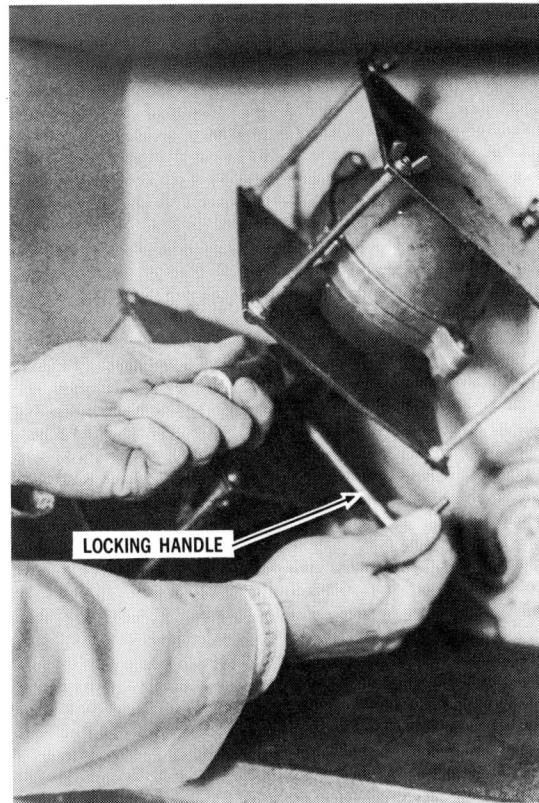

Fig. 4-10 Locking mold holders in place.

LOCKING HANDLE

Fig. 4-11 Removing mold holder.

Fig. 4-12 Loosening mold holder clamps.

Rotational Molding

24. Remove the products from the mold halves by hand. If the products stick, heat them and the mold halves in an oven. Heat them for a short time at 200° F. Then, pull the products from the mold halves by hand. To prevent sticking the next time, use a mold release, clean the mold, reduce the oven temperature, increase mold draft, or use a shorter heating cycle.

25. Check the product for bubbles or a rough inside finish. These defects can be avoided by using a smaller charge, a very low-density plastic, or a longer heating cycle. Drying the mold or raising the oven temperature may also help.

26. Check the product for brittleness. Use a very low-density plastic, raise the oven temperature, or use a longer heating cycle to avoid this problem.

27. Check the product for warpage. If it is warped, air cool the part before water cooling it or use a lower oven temperature. Cooling the mold slower, or revolving the mold while cooling it, may also help.

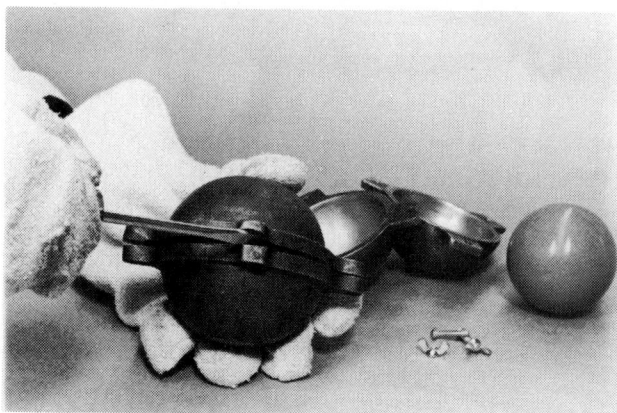

Fig. 4-13 Prying molds open.

28. Check the product for plastic webs in the narrow mold areas. Avoid this defect by raising the RPM or using a very low-density plastic.

29. Check the product for burn. Burns can be avoided by using a lower oven temperature or a shorter heating cycle. Burns are also caused by a contaminated mold. Clean the mold before using it.

30. Check the time required to mold the product. If it is too long, increase the oven temperature, use a thin-walled mold, or use a very low-density plastic.

31. Trim each product carefully with a woodworking scraper. See Fig. 4-14. Be careful not to cut yourself. The product is now completed.

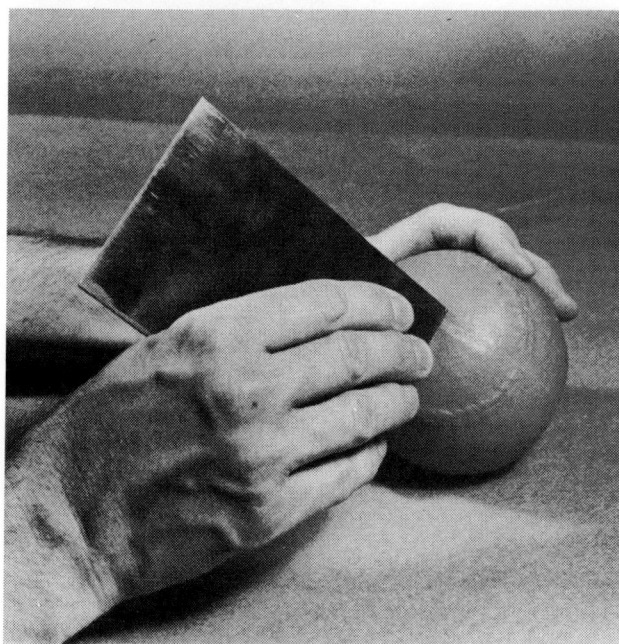

Fig. 4-14 Trimming off excess plastic.

Section II: Molding

CONCLUSION

Many rotationally molded products are shown in Fig. 4-15. Use different molds and follow the procedure just described to make these items.

Rotational molding is used to make many fully or partly closed products. Most of the products are stress-free because each product is molded without any outside pressure. Also, the design possibilities of rotational molded products are almost limitless.

REVIEW

1. List four products produced by rotational molding.

2. Describe how rotational molding works.

3. Name the type of plastic used in this activity.

4. Name the material from which your rotational mold was constructed.

5. Describe how the product wall thickness is controlled.

6. Explain the term *biaxial*.

7. List and explain each step of the rotational molding cycle.

SELECTED BIBLIOGRAPHY

Agranoff, Joan, ed. *Modern Plastics Encyclopedia*. New York: McGraw-Hill Book Company, Inc., 1976-77.

Baird, Ronald J. *Industrial Plastics*. South Holland, Illinois: The Goodheart-Willcox Company, Inc., 1971.

Bruins, Paul. *Basic Principles of Rotational Molding*. New York: Gordon & Breech, 1972.

Fig. 4-15 Examples of rotational molded products.

Milby, Robert V. *Plastics Technology.* New York: McGraw-Hill Book Company, 1973.

Patton, William J. *Plastics Technology: Theory, Design, and Manufacture.* Reston, Virginia: Reston Publishing Company, Inc., 1976.

Richardson, Terry A. *Modern Industrial Plastics.* Indianapolis: Howard W. Sams & Company, Inc., 1974.

Rotational Molding—Teacher's Manual. Indianapolis: Howard W. Sams & Company, Inc., 1974.

EQUIPMENT AND MATERIAL SUPPLIERS

1. Brodhead-Garrett, 4560 East 71st Street, Cleveland, Ohio 44105.

2. Delvies Plastics, Inc., 2320 South West Temple, P.O. Box 1415, Salt Lake City, Utah 84110.

3. Graves-Humphreys, Inc., 1948 Franklin Road, P.O. Box 1347, Roanoke, Virginia 24033.

4. Industrial Arts Supply Company, 5724 West 36th Street, Minneapolis, Minnesota 55416.

5. McKilligan Supply Corporation, 494 Chenango Street, Binghamton, New York 13901.

6. Paxton/Patterson, 5719 West 65th Street, Chicago, Illinois 60638.

The plastic, molds, and rotational molding machine needed for this assignment can be purchased from any of the suppliers listed above.

Assignment 5
Compression Molding

OBJECTIVES

To select, clean, and charge a compression mold.

To load and adjust a compression press.

To produce a compression molded product.

INTRODUCTION

Compression molding was one of the first industrial methods of molding plastics. The plastics used are thermosetting types such as phenolics, ureas, melamines, epoxies, polyesters, silicones, alkyds, and diallyl phthalates.

To make a compression molded product, a thermosetting plastic and a mold are needed. The plastic is called the "molding compound" or the "charge." Top and bottom mold halves are used. The bottom half is like the outside of the finished product. The top half is like the inside of the finished product.

To begin compression molding, the mold must be heated. The charge is placed in the hot, open mold cavity. Then, the mold is closed under pressure in a press (2,000 to 5,000 psi). This causes the charge to melt and flow all through the mold cavity. The hot charge takes the shape of the finished product.

The charge is kept hot and under pressure for a short time. This is done to cure the plastic in the mold. During curing, the charge changes chemically (crosslinking) from a powder (preform) into a hard plastic product. The mold halves are now opened and the finished product is removed.

SAFETY

The following precautions should be taken when compression molding:

1. Work in a well-ventilated area. Do not breathe plastic fumes because some are toxic.

2. Work on a heat-resistant surface.

3. Wear safety glasses and heat-resistant gloves.

4. Keep a general purpose fire extinguisher in the work area.

5. Do not touch hot plastics, molds, or machine parts with your bare hands.

6. Learn the safe operation of a compression molding press.

7. Make sure all machine control, mold, and moving part guards are in place.

8. Do not pinch your hands or fingers between the mold halves or machine parts.

9. Keep the mold halves in line and in the center of the press platens. This prevents overstressing the press and mold.

EQUIPMENT AND MATERIALS

The equipment and materials needed for compression molding are shown in Fig. 5-1. They are:

1. Safety glasses.
2. Heat-resistant gloves.
3. Heat-resistant surface.
4. Compression press.
5. Compression mold.
6. General purpose phenolic, urea, or melamine.
7. Mold release agent.
8. Cloth or paper towels.
9. Scoop.

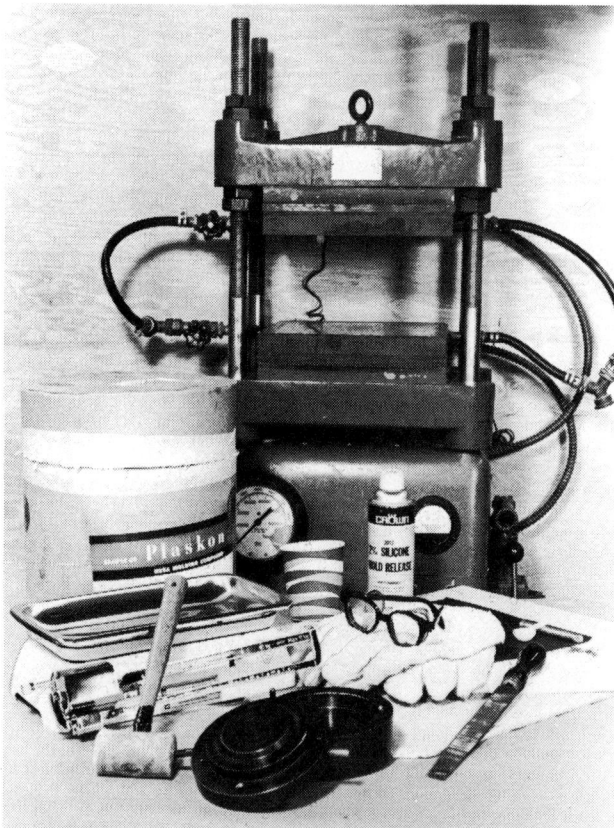

Fig. 5-1 Compression molding equipment and materials.

Section II: Molding

Fig. 5-2 Cleaning compression mold cavity.

TOP PRESS PLATEN

TOP MOLD HALF

BOTTOM MOLD HALF   BOTTOM PRESS PLATEN

Fig. 5-3 Locating bottom mold half.

10. Brass ejection rod.
11. Wood or leather mallet.
12. 350-400 grit wet/dry sandpaper.
13. Weight scales and weights.
14. 16 oz. container.
15. Single cut file.
16. Buffing machine.

COMPRESSION MOLDING LAB PROCEDURE

1. Obtain all the items listed under "Equipment and Materials" needed to compression mold a project. See Fig. 5-1.

2. Look at the compression molding press. Make sure you know where all of the press controls are located.

3. Clean the compression mold cavity (coaster mold). See Fig. 5-2.

4. Screw the top mold half into the center of the top press platen.

5. Place the bottom mold half on the bottom press platen. **Make sure it is in line with the top mold half.** See Fig. 5-3.

6. Pump the hydraulic press platens together until the mold halves touch. See Fig. 5-4. Keeping the mold halves in the platen centers prevents stressing them and the press.

7. Turn on the heater switches for the upper and lower press platens. Make sure the instructor has set the platen heater thermostats. These thermostats must be adjusted to the operating temperature of the plastic to be molded.

8. Heat the platens and mold to the temperature recommended by the plastic manufacturer. See Fig. 5-5. Urea molding compound has a temperature range of 275° F. to 325° F.

9. Check the press temperature

Compression Molding

Fig. 5-4 Pumping platens together.

Fig. 5-5 Heating the platens and mold.

gauge every five minutes. It may take 20 to 30 minutes for the mold to heat to the proper temperature. Keep the mold at the proper temperature during the following steps.

10. Open the mold by pulling the platen release lever forward.

11. Remove the lower mold from the press. See Fig. 5-6. Handle the hot mold with heat-resistant gloves. *Be careful not to burn your arms on the hot press platens.*

12. Apply a release agent to the mold halves.

13. Weigh the amount of plastic charge indicated by your instructor.

14. Place the charge in the mold cavity. See Fig. 5-7. *Do not burn your fingers on the mold.*

15. Line up the charged bottom mold half with the top mold half. Fig. 5-8.

16. Pump the mold halves together until a small amount of resistance is felt. See Fig. 5-9. **Make sure the pressure gauge reads zero.**

17. Wait 30 seconds. Open the mold slightly by pulling the platen release lever forward. (This is a mold breathing, or degassing, step.) The trapped gas can now escape.

18. Pump the mold halves together until the correct pressure is read on the pressure gauge. The instructor will tell you the correct pressure gauge reading. This gauge reading is 28,000 lbs. The area of the mold surface is 4″ and the plastic is molded at 7,000 psi. Thus 4″ x 7,000 psi = 28,000 lbs. reading on the gauge. Urea molding compound has a molding range of 2,000 to 8,000 psi.

19. Keep the mold at this pressure

Section II: Molding

Fig. 5-6   Removing the lower mold.

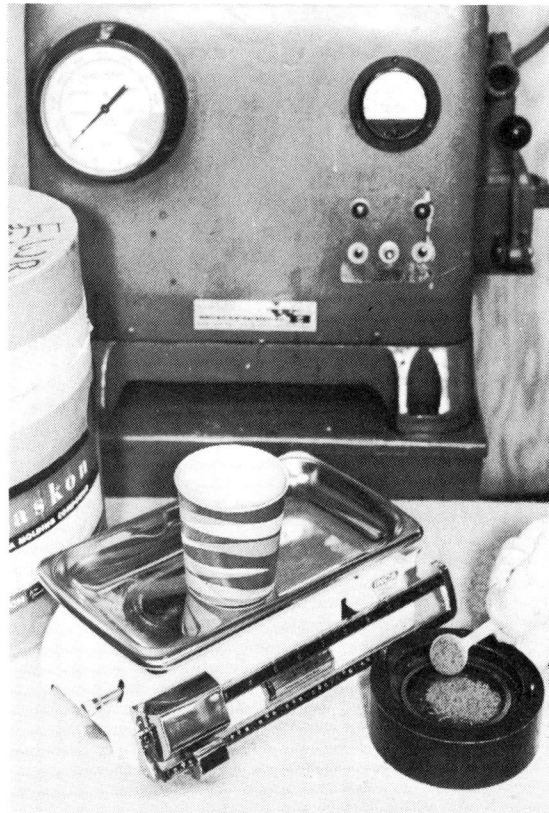

Fig. 5-7   Placing charge in mold cavity.

Fig. 5-8   Lining up mold halves.

Fig. 5-9   Pumping the mold halves together.

Compression Molding

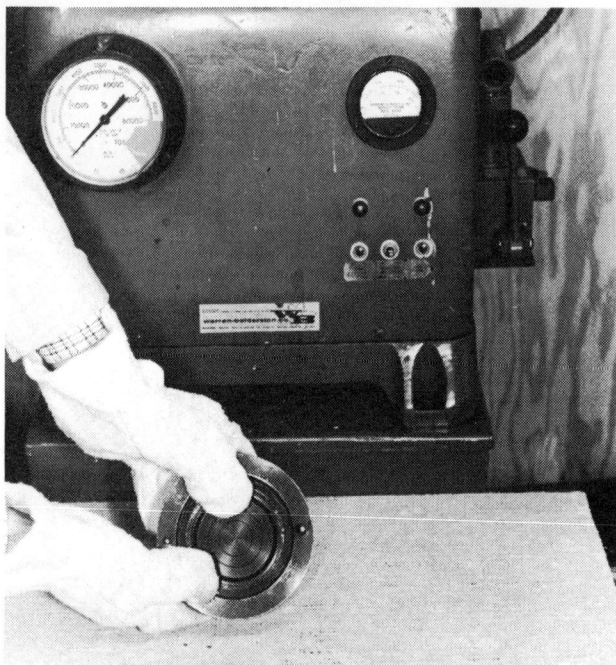

Fig. 5-10   Removing the lower mold.

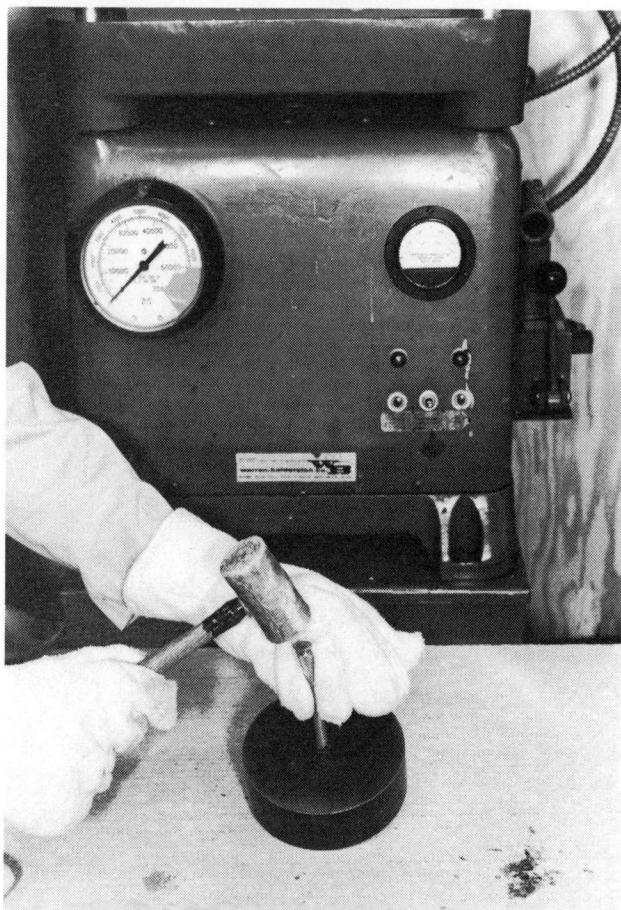

Fig. 5-11   Forcing the product from the mold.

for 4 to 5 minutes. This is the product curing stage or cycle.

20. Open the press and remove the lower mold from the platen. See Fig. 5-10.

21. Turn the mold upside down on a heat-resistant surface.

22. Remove the product from the mold. This is done by placing a brass rod on the end of the ejector pin. Hit the brass rod with a mallet to force the product out of the mold. See Fig. 5-11.

23. Break the excess plastic, or *flash,* away from the product. Use heat-resistant gloves for this process.

24. If desired, the product mold parting line can be removed. This is done by single cut filing away the line. Then, sand it with a fine wet-dry abrasive paper and buff the sanded line.

Several flash-type compression molds are shown in Fig. 5-12. Compression molds are made from carbon and alloy steels. These materials can take the high pressure produced during compression molding.

Many compression molded products are shown in Fig. 5-13. Use the procedure just described to produce these items.

CONCLUSION

Compression molding is used to rapidly produce small and medium sized products for industry and the home. A few of the home-use products include: plates, cups, saucers, pot handles, knobs, electrical wall and switch plates, ash trays, bowls, trays, coasters, and toys such as the yo-yo and checkers.

After compression molded products

are made, their flash must be removed. Industry does this by placing the parts into a barrel. The barrel has many small wooden blocks in it. The barrel is rotated. This causes the parts and wooden blocks to rub against each other. The rubbing action removes the flash and lightly polishes each plastic part.

Another industrial method for removing flash from parts is to pressure blast them. The parts are tumbled and blasted at the same time with high speed pellets (crushed fruit pits). This removes the flash and lightly polishes each part.

The flash from each part must be disposed of. If the flash is a thermoset, it cannot be remelted. It is often ground into a powder. This recycled powder (filler) is remixed in small quantities with new plastic. This mix can be recharged into compression molds.

REVIEW

1. Define *compression molding.*

2. Name the type of plastic used in this activity.

3. Describe one way of using thermosetting scrap produced during the compression molding process.

4. List four products made by compression molding.

5. Name a material from which compression molds are made.

6. List and explain each step for compression molding.

7. Define *crosslinking.*

8. Explain one way to remove flash from compression molded products.

9. Explain why you should "breathe" the compression mold.

Fig. 5-12 Examples of flash-type compression molds.

Fig. 5-13 Examples of compression molded products.

Compression Molding

SELECTED BIBLIOGRAPHY

Agranoff, Joan, ed. *Modern Plastics Encyclopedia*. New York: McGraw-Hill Book Company, 1976-77.

Baird, Ronald J. *Industrial Plastics*. South Holland, Illinois: The Goodheart-Willcox Company Inc., 1971.

Compression and Transfer Molding-Teacher's Manual. Indianapolis: Howard W. Sams & Company, Inc., 1974.

Milby, Robert V. *Plastics Technology*. New York: McGraw-Hill Book Company, 1973.

Patton, William J. *Plastics Technology: Theory, Design, and Manufacture*. Reston, Virginia: Reston Publishing Company, Inc. 1976.

Richardson, Terry A. *Modern Industrial Plastics*. Indianapolis: Howard W. Sams & Company, Inc., 1974.

Rosato, Dominick V., ed. *Plastics Industry Safety Handbook*. Boston: Cahners Books, Inc., 1973.

Vaill E. W. *Molding Thermoset Materials*. New York, New York: Society of Plastics Industry, Inc., 1970.

EQUIPMENT AND MATERIAL SUPPLIERS

1. Brodhead-Garrett, 4560 East 71st Street, Cleveland, Ohio 44105.

2. Cope Plastics, Inc., 4441 Industrial Drive, Godfrey, Illinois 62035.

3. Delvie's Plastics, Inc., 2320 South West Temple, P.O. Box 1415, Salt Lake City, Utah 84110.

Section II: Molding

4. Graves-Humphreys, Inc., 1948 Franklin Road, P.O. Box 1347, Roanoke, Virginia 24033.

5. Industrial Arts Supply Company, 5724 W. 36th St., Minneapolis, Minnesota 55408.

6. McKilligan Industrial Supply Corporation, 494 Chenango Street, Binghamton, New York 13901.

7. Paxton/Patterson, 5719 West 65th Street, Chicago, Illinois 60638.

8. Vicor Plastic Equipment, Inc., 231 E. 1st Avenue, Roselle, New Jersey 07203.

All of the equipment and materials needed for compression molding can be purchased from any of the suppliers listed above.

Compression Molding

Assignment 6
Blow Molding

OBJECTIVES

To adjust the blow molding equipment.

To make a blow molded product.

INTRODUCTION

Blow molding is a mass production industrial process. Hollow products, such as small pill bottles and large 55 gallon drums, are made with this process. The blow molding process was first used in the glass industry to make containers. A large, commercial blow molding installation is shown in Fig. 6-1.

Fig. 6-1 Commercial blow molding installation.   (Courtesy Voith Fischer Plastics Machines, Inc.)

Section II: Molding

Many plastic products are also blow molded. A good example of a hollow, blow-molded toy is the all plastic tricycle. Because of the hollow body and frame, a small child is able to carry it. Because it is made from very durable plastic, it is extremely strong and light weight.

Blow molding is done by using a machine called a *blow molder*. A special blow molding plastic is place into the extruder hopper. The extruder is heated to the operating temperature and turned on. The motor rotates an extruder screw. This screw mixes and forces the hot plastic through a die head. From the die head, the hot plastic is pushed downward through a die. The die forms the plastic into the shape of a tube called a "parison."

The parison is then extruded between the mold halves. The mold is closed. This closing pinches and welds the bottom of the parison together. Next, air is blown into the top of the parison. This causes it to expand to the shape of the mold. The mold is quickly cooled. After cooling, the mold is opened and the product is removed. The product is now trimmed, printed and packaged.

SAFETY

The following precautions should be taken when blow molding:

1. Work in a well-ventilated area. Do not breathe plastic fumes because some are toxic.

2. Work on a heat-resistant surface.

3. Wear safety glasses and heat-resistant gloves.

4. Keep a general purpose fire extinguisher in the work area.

5. Do not touch the hot plastic

parison, extruder barrel, die head, and die.

6. Learn the safe operation of the blow molding equipment.

7. Do not touch the electric heater wires.

8. Make sure all machine control and part guards are in place and working.

9. Do not place your hands, face, or body in front of the die.

10. Keep your arms and hands out of the hopper.

11. Keep your fingers, hands, and arms away from the clamps or mold halves.

12. Use the air hose only as directed.

EQUIPMENT AND MATERIALS

The equipment and materials needed for blow molding are:

1. Safety glasses.
2. Heat-resistant gloves.
3. Heat-resistant surface.
4. Blow molder.
5. Extruder.
6. Air lines.
7. Mold.
8. PVC or polyethylene.
9. Utility knife.
10. Paper or metal cutting shears.

BLOW MOLDING LAB PROCEDURE

This experiment will be done on a small, school laboratory extrusion blow molding machine, as in Fig. 6-2.

1. Ask your instructor if each zone heater has been adjusted to the correct temperature. For medium

Fig. 6-2 School laboratory extrusion blow molding machine.

Section II: Molding

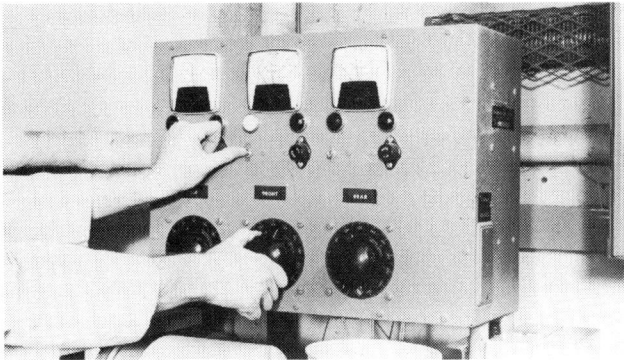

Fig. 6-3   Turning on zone heater switches.

Fig. 6-4   Heat gauges and lights.

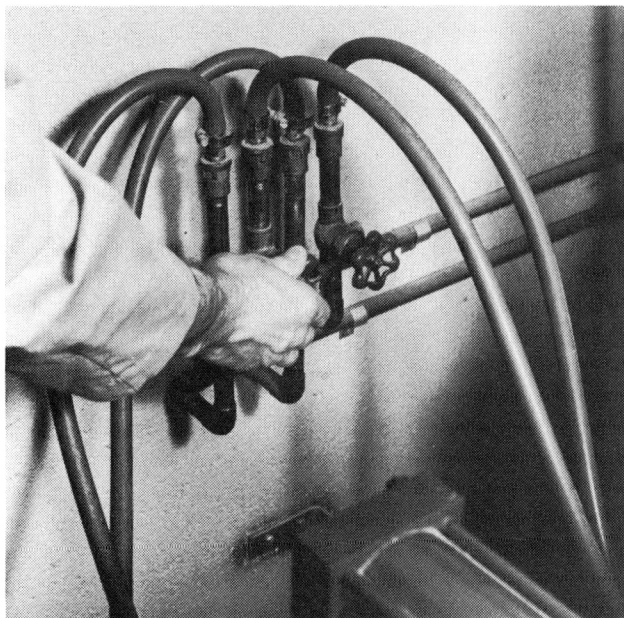

Fig. 6-5   Turning on cooling water.

density polyethylene, one manufacturer recommends the following zone heater thermostat settings: rear zone equals 300° F., front zone equals 325° F., and the die equals 325° F.

2. Turn on each zone heater switch. See Fig. 6-3 (left hand).

3. Turn each zone heater transformer knob three quarters of a turn. See Fig. 6-3 (right hand).

4. Let the machine reach the temperature setting. Watch for all of the zone gauge heater lights to switch from red to green. See Fig. 6-4.

5. Allow the machine to heat at these temperatures for 15 to 20 minutes. This gives the inside machine parts (screw and barrel) time to completely heat. **Do not turn on the machine screw before the warm up period is over.** If the screw is turned on too early, it may break. It could also scar the machine barrel. Be sure to ask your instructor before turning the machine on.

6. Make sure your instructor has adjusted the blow mold closed clamp timer(s). One hundred seconds were set for this product.

7. Watch for soft, clear plastic to appear at the die opening.

8. Fill the hopper with plastic.

9. Turn on the blow mold cooling water. See Fig. 6-5.

10. Turn on the extruder. See Fig. 6-6. **Be sure to ask the instructor before turning on the extruder.**

11. Extrude a parison between the mold halves, as in Fig. 6-7. The parison should extend a little beyond the mold pinch off plate. The parison should be straight. It

Blow Molding

Fig. 6-6 Turning on extruder.

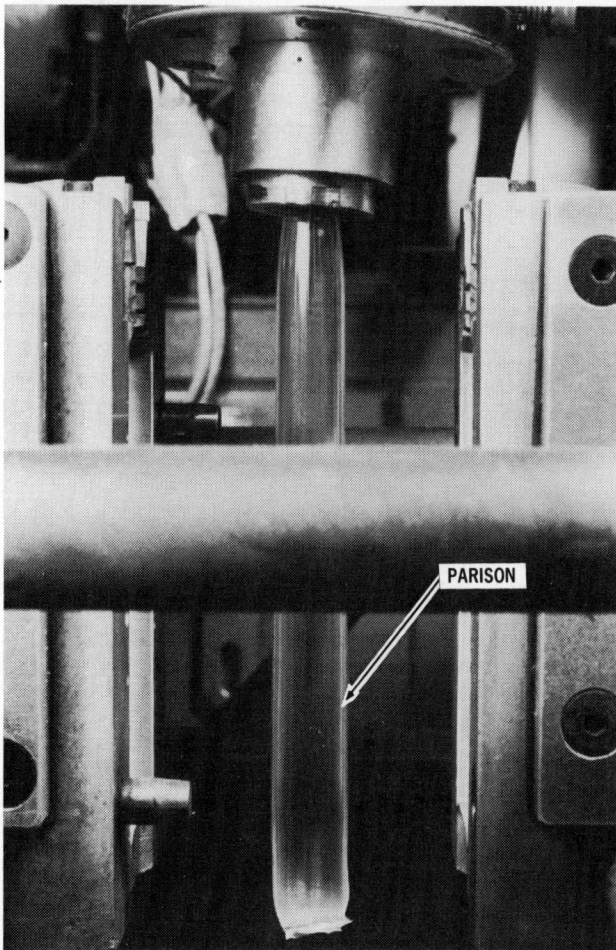

Fig. 6-7 Extruding parison between mold halves.

should have an equal wall thickness.

12. Turn off the extruder.

13. Push the mold clamp close switch and let the mold close.

14. Turn on the blow molding air control. See Fig. 6-8. The air pressure is 10 to 60 psi. (Ask your instructor to tell you what air pressure to use.) Air flows into the parison from the neck area of the mold. The product is formed at this time.

15. Turn off the blow molding air.

16. Wait for the mold to automatically open.

17. Wear heat-resistant gloves and remove the product by pulling it away from the mold neck area. See Fig. 6-9. The mold is warm and the ends of the plastic may be hot.

18. Trim off the extra pinch-off plastic at the bottom of the product. See Fig. 6-10. The mark left on the bottom of the product is called a "scar" (pinch-off).

19. Trim off the extra plastic at the neck of the product. Use the cutting equipment recommended by your instructor.

Fig. 6-11 shows the completed bottle. It is now ready for filling and capping. Many products that were blow molded are shown in Fig. 6-12. Each product requires its own mold.

CONCLUSION

The process just shown is called *extrusion blow molding*. It is only one type of blow molding. *Injection* and *preform* are two other blow molding processes.

Fig. 6-8   Turning on blow molding air control.

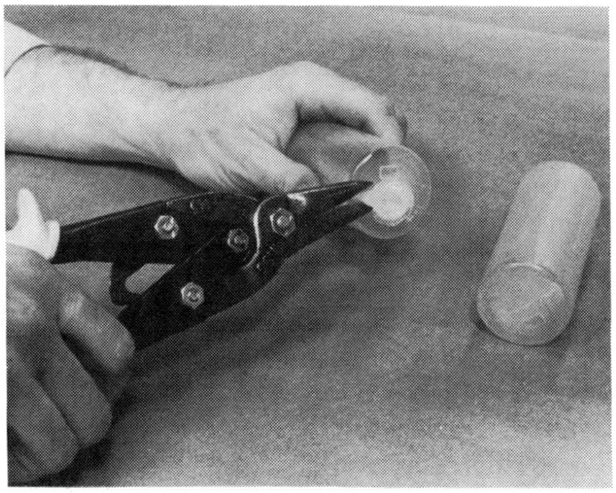

Fig. 6-10   Trimming excess plastic from bottom.

Fig. 6-9   Removing product.

Fig. 6-11   Completed bottle.

Fig. 6-12   Examples of blow molded products.

63   Blow Molding

In injection blow molding, the injection machine makes the parison. Injection blow molding products require very little trimming. Also, irregular shaped products may be formed with this process.

Preform blow molding is a process in which a parison is formed, cooled, cut to length, and stored. Later the parison is softened in an oven. It is then placed in a mold and blown into a product.

The materials blow molded are thermoplastics. Any clean scraps produced may be reused. Reusing this material is done as was described in Assignment 3 (Injection Molding).

REVIEW

1. Define *blow molding.*

2. Explain each step in extrusion blow molding.

3. Name the type of plastic used in this activity.

4. Describe how to reuse thermoplastic scrap produced during the blow molding process.

5. List three products made by the blow molding process.

6. Describe how to identify blow molded products.

7. Describe the preform blow molding process.

8. Name two advantages of making a product by injection blow molding instead of extrusion blow molding.

9. Define *parison.*

SELECTED BIBLIOGRAPHY

Agranoff, Joan, ed. *Modern Plastics Encyclopedia.* New York: McGraw-Hill Book Company, 1976-77.

Section II: Molding

Baird, Ronald J. *Industrial Plastics.* South Holland, Illinois: The Goodheart-Willcox Company, Inc., 1971.

Extrusion Blow Molding-Teacher's Manual. Indianapolis: Howard W. Sams & Company, Inc., 1974.

Milby, Robert V. *Plastics Technology.* New York: McGraw-Hill Book Company, 1973.

Patton, William J. *Plastics Technology: Theory, Design, and Manufacture.* Reston, Virginia: Reston Publishing Company, Inc., 1976.

Richardson, Terry A. *Modern Industrial Plastics.* Indianapolis: Howard W. Sams & Company, Inc., 1974.

Rosato, Dominick V., ed. *Plastics Industry Safety Handbook.* Boston: Cahners Books, 1973.

EQUIPMENT AND MATERIAL SUPPLIERS

1. Brodhead-Garrett, 4560 East 71st Street, Cleveland, Ohio 44105.

2. Brown Plastics Engineering Company, Inc., 1823 Holste Road, Northbrook, Illinois 60062.

3. Cope Plastics, Inc., 4441 Industrial Drive, Godfrey, Illinois 62035.

4. Graves-Humphreys, Inc., P.O. Box 13407, 1948 Franklin Road, Roanoke, Virginia 24033.

5. Paxton/Patterson, 5719 West 65th Street, Chicago, Illinois 60638.

The blow molder can be purchased from any of the suppliers listed above. Blow molding plastics and molds are available from suppliers 1, 3, 4, and 5.

Assignment 7
Extrusion

OBJECTIVES

To adjust the extruder and take-off equipment.

To extrude a profile shape or part.

INTRODUCTION

Extrusion is a high volume industrial process. With this process, small (lacing) and large (drain pipes) products are made. The products are either solid or hollow. The parts often have the same cross-sectional size throughout their lengths.

Many plastics products are extruded. A few such products include rod, pipe, tubing, sheet, coated wire, film, weather stripping, screwdriver handles, and coated paper. Each product is called an "extrudate" as it leaves the extruder.

The extrusion process works as shown in Fig. 7-1. Thermoplastic in pellet or powder form is placed in the extruder hopper. The extruder is heated to the operating temperature and turned on. The motor rotates an extruder screw.

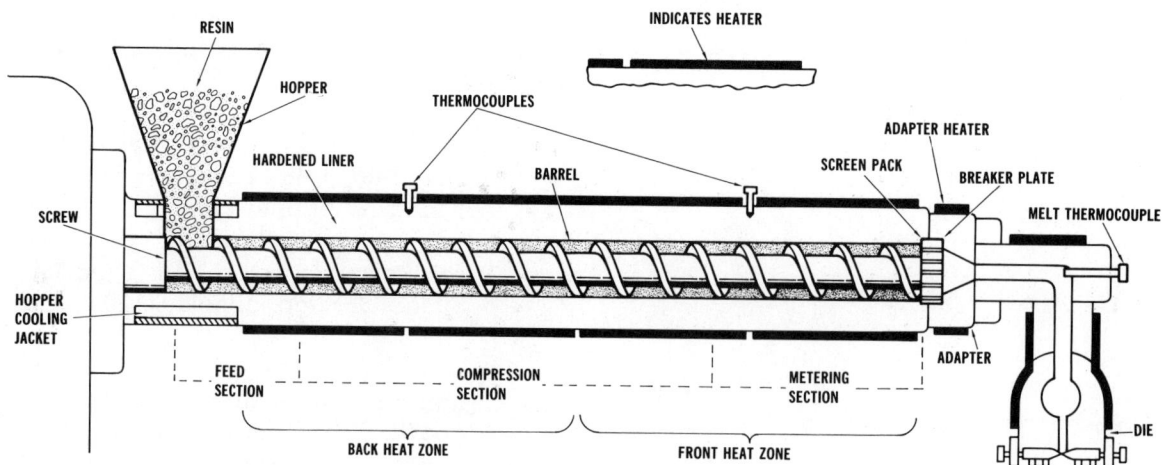

Fig. 7-1 The extrusion process. (Courtesy Howard W. Sams & Co., Inc.)

Section II: Molding

This screw mixes and forces the hot plastic through a die. The die forms the plastic into the shape of the product. (A cross-sectional shape of the product is called a "profile.") From the extruder die, the plastic goes to the take-off equipment. Take-off equipment sizes, cools, winds, and cuts the product for storage. Fig- 7-2 shows a large industrial extruder setup.

SAFETY

The following precautions should be taken when extruding:

1. Work in a well-ventilated area. Do not breathe plastic fumes because some are toxic.

2. Work on a heat-resistant surface.

Fig. 7-2   Industrial extruder set-up. (Courtesy Egan Machinery Company)

Extrusion

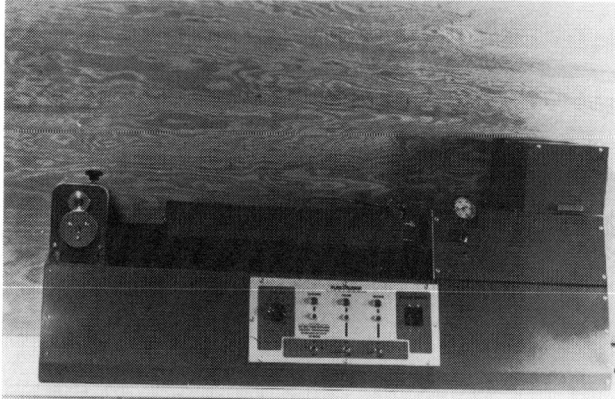

Fig. 7-3 School laboratory extruder. (Courtesy Brodhead-Garrett Company)

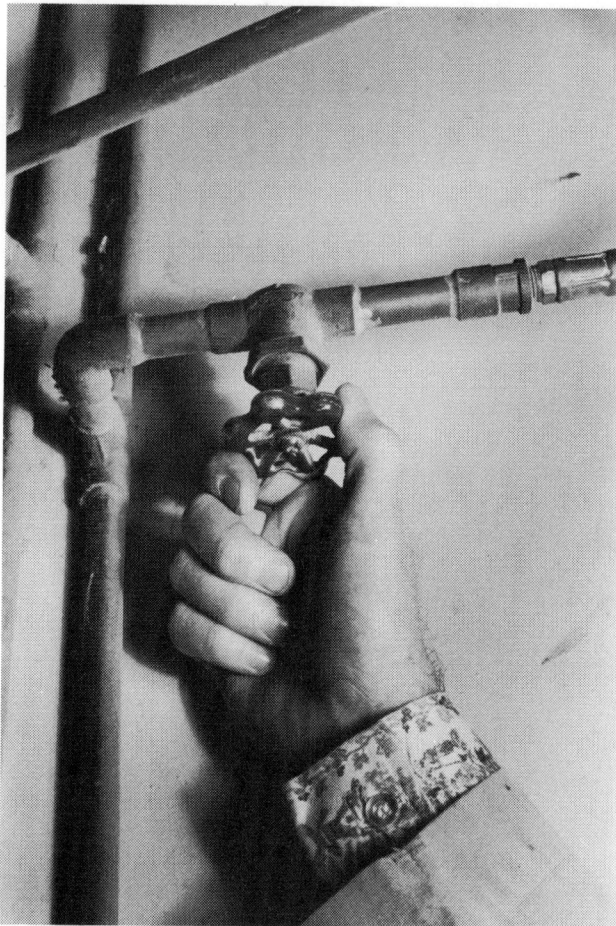

Fig. 7-4 Turning on main water cooling valve.

3. Wear safety glasses and heat-resistant gloves.

4. Keep a general purpose fire extinguisher in the work area.

5. Do not touch the hot plastic extrudate, extruder barrel, or die.

6. Learn the safe operation of the extrusion equipment.

7. Make sure all machine control and moving part guards are in place and working.

8. Do not place your hands, face, or body in front of the die.

9. Do not touch the electric heater wires.

10. Keep your hands and arms out of the hopper.

11. Keep your hands and arms out of the puller.

EQUIPMENT AND MATERIALS

The equipment and materials needed for the extrusion process are:

1. Safety glasses.
2. Heat-resistant gloves.
3. Heat-resistant surface.
4. Extruder.
5. Water tank.
6. Puller.
7. Extruder die. (Rod, tube, ribbon, lacing, or profile.)
8. Polyethylene, PVC, or polystyrene.
9. Scoop.
10. Sizer plate.
11. Gasket.
12. 2 or 3 hold-down bars.
13. Brass, wood, or aluminum threading tool.
14. Metal cutting shears.

Section II: Molding

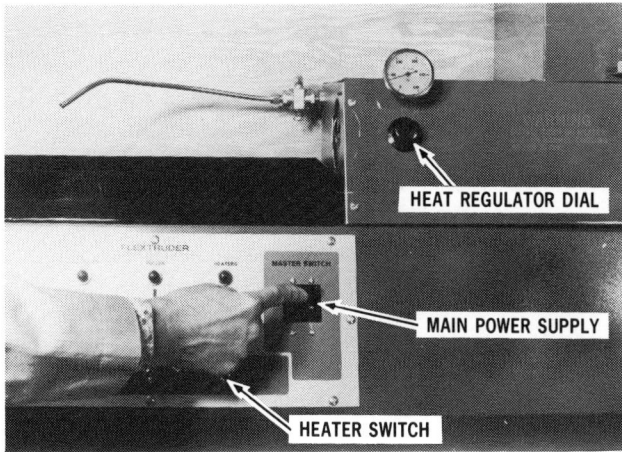

Fig. 7-5 Main power supply, heater switch, and heat regulator dial.

Fig. 7-6 Placing water tank in extruder reservoir.

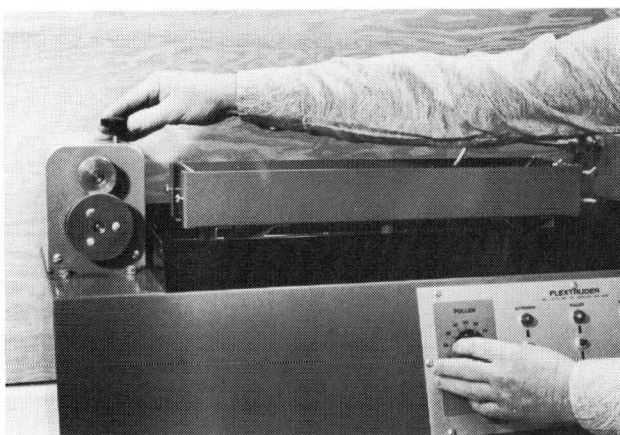

Fig. 7-7 Adjusting gap between puller rolls.

EXTRUSION LAB PROCEDURE

The extrusion procedure described here will be done on a small, school laboratory extruder. See Fig. 7-3.

1. Turn on the extruder main water cooling valve. See Fig. 7-4. Make sure water is flowing through all cooling lines.

2. Switch on the extruder main power supply. See Fig. 7-5.

3. Switch on the extruder heaters.

4. Adjust the heat regulator dial. This adjustment will be different for each type of plastic extruded. Ask your instructor to give you the proper setting. A setting of 60 was used for this operation. This setting gave a 285° F. operating temperature for the powdered polyethylene.

5. Let the extruder heat at this temperature for 30 minutes. This gives the machine (screw and barrel) time to completely heat. **Do not turn on the extruder before the warm up period is over.** If the screw is turned on too soon, it may break. It could also scar the barrel. Be sure to check with your instructor before turning on the extruder.

6. Place the water tank in the extruder reservoir. Make sure it is pushed toward the puller end of the reservoir. See Fig. 7-6.

7. Fill the hopper with plastic. Do this only after the extruder temperature has stabilized.

8. Turn on the puller switch.

9. Adjust the puller to a slow speed.

10. Adjust an ⅛" gap between the puller rolls. See Fig. 7-7.

Extrusion

Fig. 7-8 Turning on water tank filler valve.

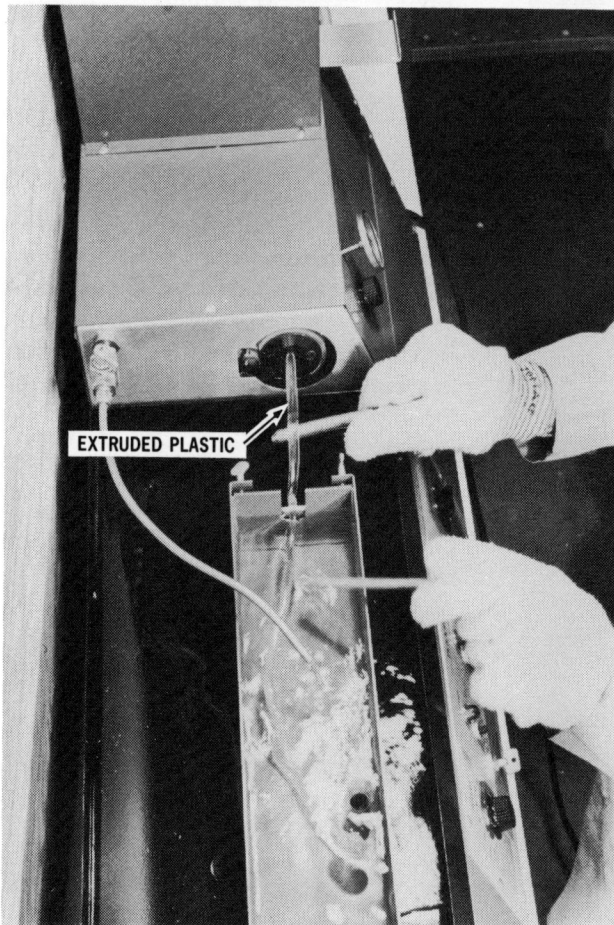

EXTRUDED PLASTIC

Fig. 7-9 Moving the hot plastic.

11. Turn on the water tank filler valve. See Fig. 7-8. The water is adjusted so that it runs out from the bottom half opening in the sizer plate.

12. Turn on the extruder motor. Be sure to ask the instructor before turning on the extruder.

13. Wear heat-resistant gloves and use a dowel, brass, or aluminum rod to move the hot plastic. *Do not touch the plastic with your fingers.* Lift the hot extruded plastic over the top of the bottom sizing plate. Make sure the plastic is placed in the water tank. Fig. 7-9.

14. Pull the extruded plastic slowly through the water tank. Keep a little tension on the extruded plastic.

15. Thread the extruded plastic through the water tank gasket and the puller rollers. If lumps, twists, or threading problems happen, snip off the plastic and repeat Steps 13, 14, and 15.

16. Adjust the puller speed to match the extrusion speed. The extrudate will thin and break if the puller runs too fast. The extrudate will sag if the puller runs too slow.

17. Put the metal hold-down bars in the water tank over the extruded plastic. See Fig. 7-10. This keeps the plastic under water and cooling.

18. Move the water tank about 2½" from the extruder die.

19. Fasten the top half of the sizer plate over the bottom half. See Fig. 7-11. You may have to run the puller fast. This will make the extrudate smaller than the sizer plate hole. **Be careful not to thin and break the extrudate.**

20. Adjust the puller speed so the plastic fills up the sizer plate hole.

Section II: Molding

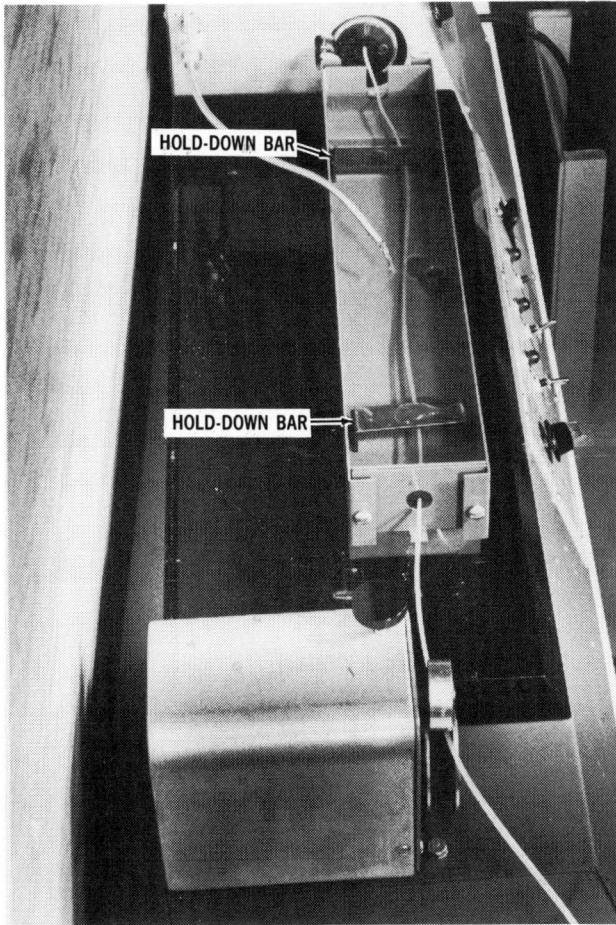

Fig. 7-10 Putting hold-down bars over plastic.

21. Check the shape and quality of the extruded part from time to time. There will be defects in the extruded part until the system stabilizes. Slow down the puller speed if the product is too thin. If it is twisted, adjust the puller rolls. A dirty product can be avoided by purging or cleaning the extruder screw and barrel. Stiff products can be avoided by increasing the barrel temperature. Lowering the barrel temperature can prevent a soft product. If the product sags at the die, speed up the puller.

22. Cut off the extruded part when the correct length has been reached. Use metal cutting shears. See Fig. 7-12.

23. Fig. 7-13 shows a coil of ⅛" polyethylene hot gas welding rod. This is the completed project.

CONCLUSION

Fig. 7-14 illustrates a number of extruded products. Some of these products are straws, pipes, tubing, gaskets, coated wire, clips, screwdriver handles, and welding rods.

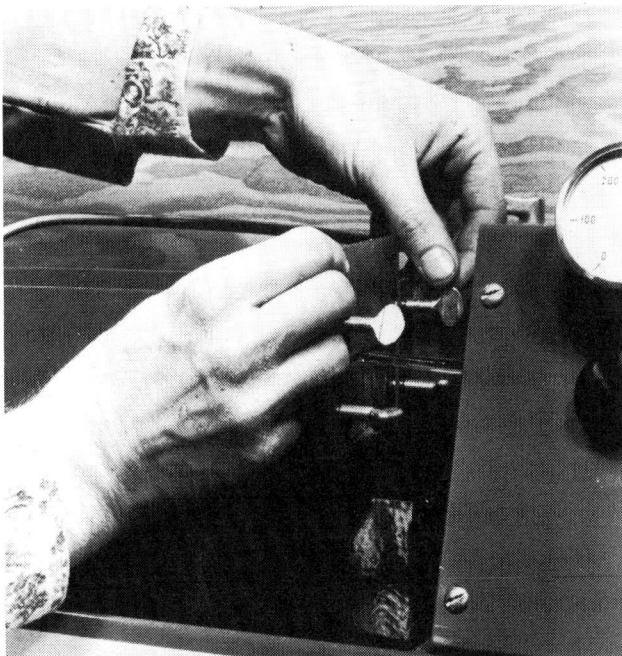

Fig. 7-11 Fastening sizer plate halves.

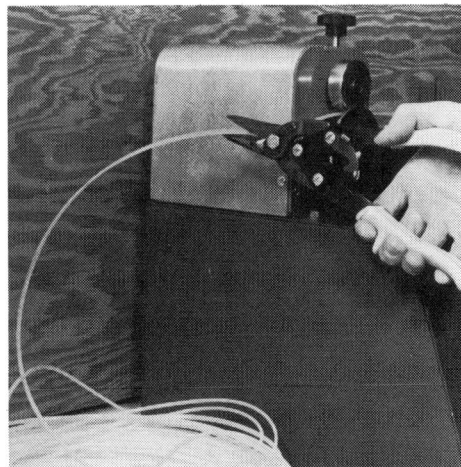

Fig. 7-12 Cutting extruded plastic.

71

Extrusion

Fig. 7-13 Coil of polyethylene hot gas welding rod.

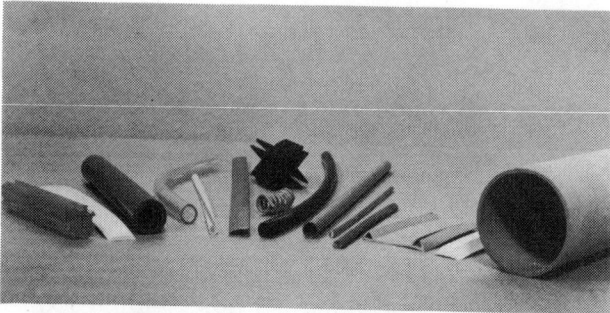

Fig. 7-14 Examples of extruded products.

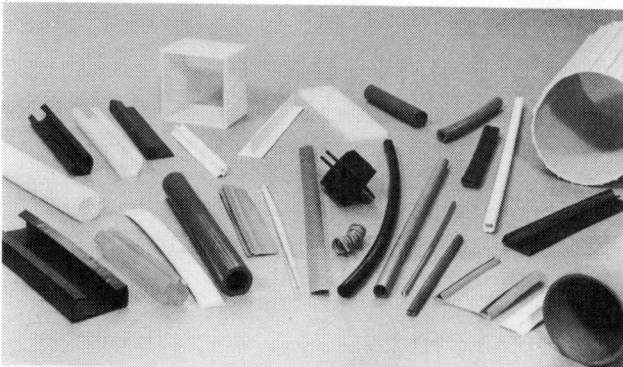

Fig. 7-15 Examples of extruded profiles.

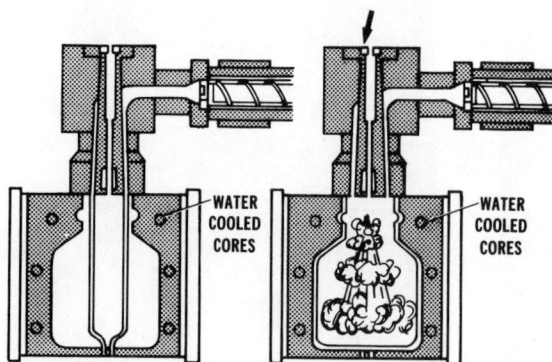

Fig. 7-16 Schematic of extrusion blow molding. (Courtesy Howard W. Sams & Co., Inc.)

Fig. 7-15 shows many extruded profiles. The procedure just described is used to make these items. Only the die and material are changed.

Often the extruder is used as a blender or mixing machine for plastics. Ingredients are placed in the hopper. The heated extruder screw is turned on and mixes the materials. As the blended plastic comes out of the die, it is chopped into small pellets. The extruder is often used for other purposes. It supplies plastic for blow molders (see Fig. 7-16), wire coaters (see Fig. 7-17), and film blowing machines.

REVIEW

1. Describe each step in the extrusion process.

2. Name the type of plastic used in this activity.

3. Define extrusion.

4. List four products made by extrusion.

5. Why is a water tank used in this activity?

6. Why is a sizer plate used in this activity?

7. Explain what a puller does.

8. List four uses for extruders.

9. Show how to adjust equipment for a sagging extrudate.

10. Describe how to adjust equipment for a twisted extrudate.

SELECTED BIBLIOGRAPHY

Agranoff, Joan, ed. Modern Plastics Encyclopedia. New York: McGraw-Hill Book Company, 1976-77.

Section II: Molding

Baird, Ronald J. *Industrial Plastics.* South Holland, Illinois: The Goodheart-Willcox Company, Inc., 1971.

Extrusion Blow Molding-Teacher's Manual. Indianapolis: Howard W. Sams & Company, Inc., 1974.

Milby, Robert V. *Plastics Technology.* New York: McGraw-Hill Book Company, 1973.

Patton, William J. *Plastics Technology: Theory, Design, and Manufacture.* Reston, Virginia: Reston Publishing Company, Inc., 1976.

Richardson, Paul N. *Introduction to Extrusion.* Greenwich, Connecticut: Society of Plastics Engineers, Inc., 1974.

Richardson, Terry A. *Modern Industrial Plastics.* Indianapolis: Howard W. Sams & Company, Inc., 1974.

Rosato, Dominick V., ed. *Plastics Industry Safety Handbook.* Boston: Cahners Books, 1973.

EQUIPMENT AND MATERIAL SUPPLIERS

1. Brodhead-Garrett, 4560 East 71st Street, Cleveland, Ohio 44105.

2. Brown Plastics Engineering Company, Inc., 1823 Holste Road, Northbrook, Illinois 60062.

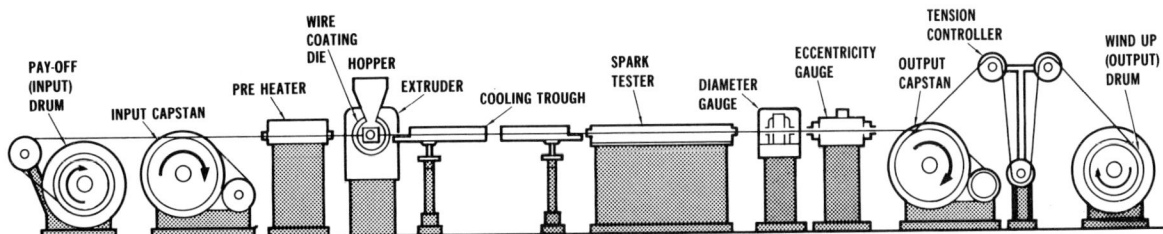

Fig. 7-17 Extrusion coating of wire and cable.

(Courtesy Howard W. Sams & Co., Inc.)

Extrusion

3. Cope Plastics, Inc., 4441 Industrial Drive, Godfrey, Illinois 62035.

4. C. W. Brabender Instruments Inc., 50 East Wesley Street, South Hackensack, New Jersey 07606.

5. McKilligan Industrial Supply Corporation, 494 Chenango Street, Binghamton, New York 13901.

6. Rainville Co., Inc., 200 Clay Avenue, Middlesex, New Jersey 08846.

7. Wayne Machine & Die Company, 100 Furler Street, Totowa, New Jersey 07512.

Extruders and molds can be purchased from any of the suppliers listed above. Plastic can be purchased from suppliers 1, 3, and 5.

Assignment 8
Thermofusion

OBJECTIVE

To design and make a thermofusion project.

INTRODUCTION

Thermofusion is a process in which thermoplastics are melted together. They are in the form of pellets, powders, lumps, or crystals.

In a typical project, the templet is placed on a mold (cookie sheet). The plastic is placed in the templet. The templet is then removed from the plastic and mold. Sometimes the templet is not removed from the mold. It becomes a part of the finished product (imitation stained glass). In other projects, the templet may not even be used. The plastics may be placed on the mold in any design. The plastic and mold are put into an oven and heated. The heat causes the plastic to melt and flow into a product. The product and mold are cooled. Then, the product is removed from the mold.

SAFETY

The following precautions should be taken when thermofusion molding:

1. Work in a well-ventilated area. Do not breathe plastic fumes because some are toxic.

2. Work on a heat-resistant surface.

3. Wear safety glasses and heat-resistant gloves.

4. Keep a general purpose fire extinguisher in the work area.

5. Do not touch hot plastics, molds, or metal surfaces with your bare hands.

Thermofusion

6. Learn the safe operation of the oven.

EQUIPMENT AND MATERIALS

The equipment and materials needed for thermofusion molding are:

1. Safety glasses.
2. Heat-resistant gloves.
3. Heat-resistant surface.
4. Vented-type industrial or laboratory oven containing a heat circulating fan.
5. Molds. (Cookie sheet, heat-resistant glass, heavy metal foil, metal containers, metal bowls, or metal pans.)
6. Ground polystyrene and polyethylene scraps (colored or clear) from product wastes. Polystyrene cooking pellets or mosaic tile are available at hobby stores.
7. Mold release agent.
8. Funnel.
9. Metal foil.
10. Pencil, sketch pad, and tracing paper (8½" x 11").
11. Light gauge metal for templets.
12. Metal cutting shears.

Fig. 8-1   Thermofusion oven and molds.

THERMOFUSION LAB PROCEDURE

1. Obtain the oven and mold(s). The molds can be cookie sheets, heat-proof glass, metal foil, metal pans, or bowls. See Fig. 8-1. Molds often give shape to the finished project. They also prevent the plastic from spilling into the oven.

2. Obtain all the items listed under "Equipment and Materials" needed for thermofusion project. See Fig. 8-2.

3. Draw the thermofusion project design full size on a sheet of paper.

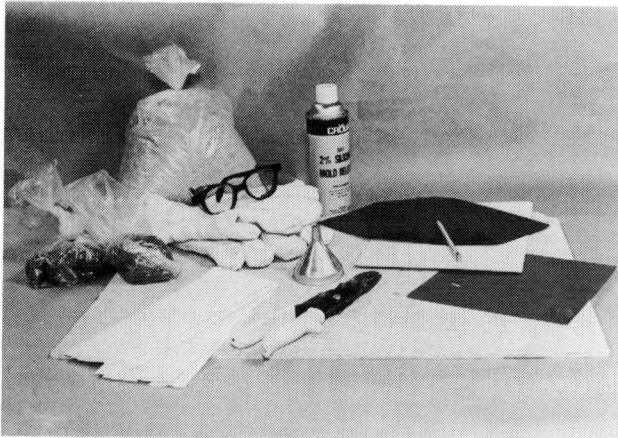

Fig. 8-2   Thermofusion tools and materials.

Section II: Molding

Fig. 8-3   Cutting design out of metal.

Fig. 8-4   Filling open areas in templet.

Fig. 8-5   Lifting templet(s) out of plastic.

4. Transfer the project design to a clean, light-gauge metal.

5. Cut the traced design out of the light gauge metal. See Fig. 8-3. Use the tools, equipment, and procedure suggested by your instructor. (The parts cut from the metal are called templets. Templets can also be made by soldering or welding strips of metal together.)

6. Apply a release agent to the mold surface.

7. Place the templets in the center of the mold. Small handles called "tabs" are soldered or glued to the templets. The tabs are used when the templets are lifted in and out of the mold.

8. Fill a funnel nearly full with thermofusion plastic. Feed the plastic from the funnel into the correct open areas in the templet. See Fig. 8-4. The depth of the plastic should be about ¼". When the plastic is heated it softens. Its depth reduces as the plastic flows together. If the plastic depth is less than ¼", the finished project may develop holes.

9. Use the templet tabs to lift the templet(s) straight out of the plastic. See Fig. 8-5.

10. Place plastic, as shown in Step 8, into the plastic-free areas. See Fig. 8-6.

11. Place the mold in a 350°F. preheated oven. Wear heat-resistant gloves when working with hot ovens and molds.

12. Check the plastic in the mold and oven every 5 to 7 minutes. If this project is smooth and shiny, it may be taken from the oven. Often a rough surface is wanted for a project. The project is then taken from the oven before the plastic has

Thermofusion

Fig. 8-6 Placing plastic in plastic-free areas.

Fig. 8-7 Heating the plastic.

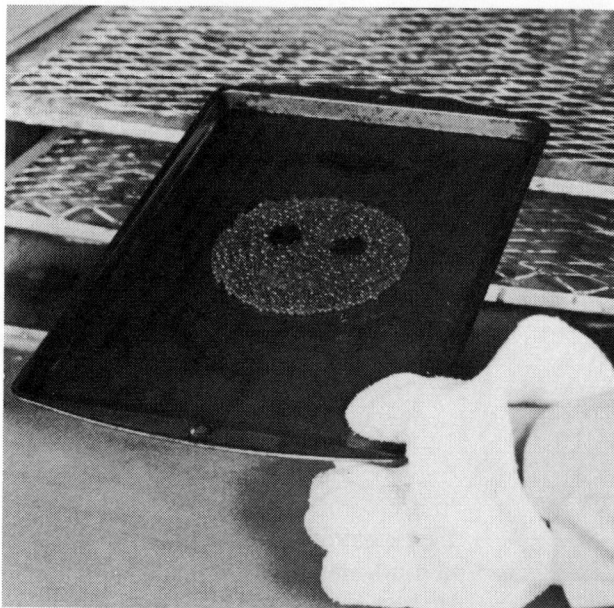

Fig. 8-8 Removing mold from oven.

heated enough to smooth it out. See Fig. 8-7.

13. Remove the mold and project from the oven. See Fig. 8-8. The project has fused into one sheet of plastic. Often, plastic is taken from the oven and mold in a hot and limp sheet form. It is then placed over a jar or can. It will then be shaped like a dish or bowl.

14. Place the mold and project on a heat-resistant surface and let them cool.

15. Remove the project from the mold. See Fig. 8-9. If the project sticks to the mold, work a thin piece of wood between the part and the mold. **Do not touch the mold with a metal tool.** A metal tool will scratch the mold surface.

16. A rough and a smooth textured finished product is shown in Fig. 8-10. If thin spots or holes are found in the project, they can be repaired. Fill the holes or thin spots with plastic pellets. Place the mold and project into the heated oven. Melt the added pellets to the plastic project.

Some thermofusion projects are made to look like stained glass. They are made by melting plastic between soldered lead strips. Projects with rippled surfaces can be made. They are made by placing crinkled metal foil in the mold and melting plastic over the foil. Designs and objects may be inserted between layers of plastic crystals. When the plastic is melted, the object is embedded in the plastic.

CONCLUSION

The thermofusion methods described here are not used by industry to mass produce products. Thermofusion is used by industry to make a small number of large

Section II: Molding

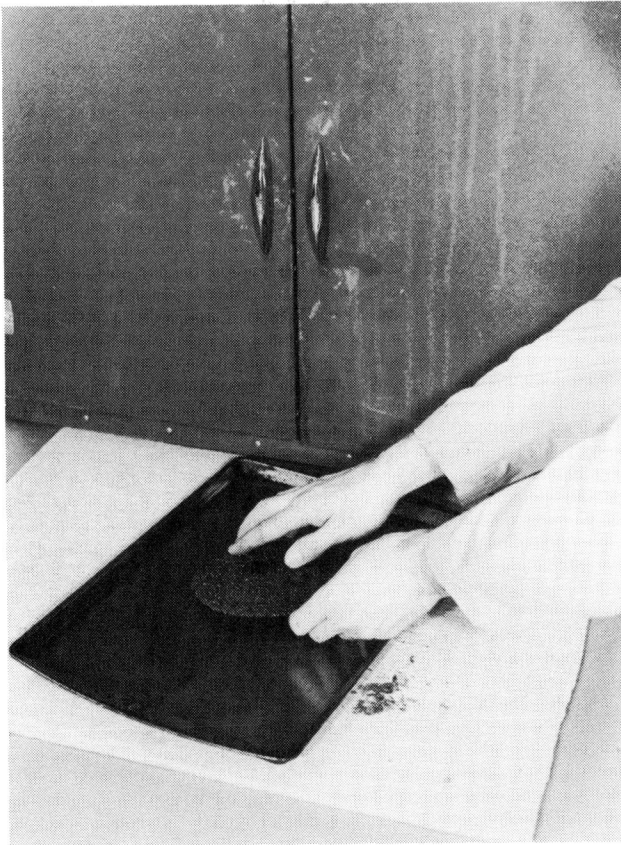

Fig. 8-9 Removing project from mold.

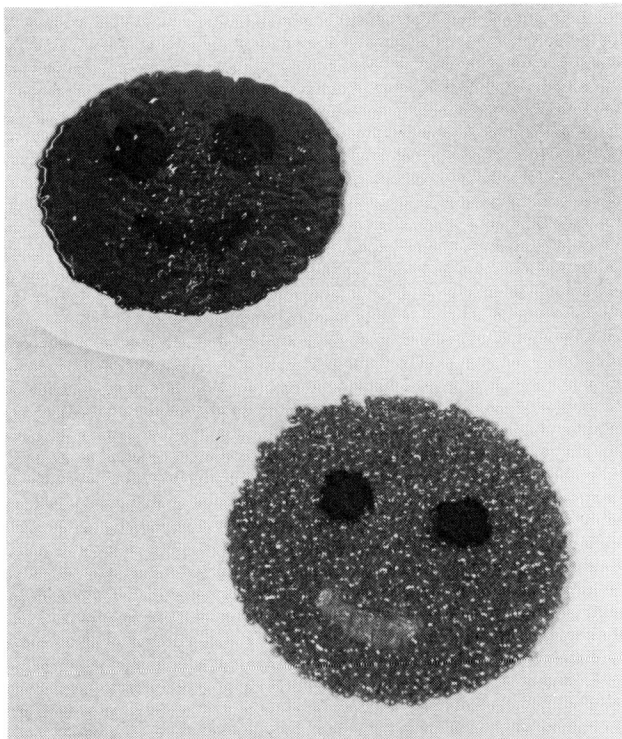

Fig. 8-10 Completed thermofusion projects.

products such as containers. This process is also used to make many hobby type products.

Another of the industrial thermofusion processes is called *static molding.* This process may be shown in the school shop by making a cup.

A metal cup mold must be used to static mold a cup. The thermoplastic is poured into the mold through its open top. When the cup mold is filled with thermoplastic, it is covered with a heatproof lid. The mold is then heated. This heat causes the plastic next to the mold walls to melt together. This forms the outside of the cup. The heatproof lid is then removed and the plastic that did not melt is poured out.

The cup mold with its plastic lining is heated again, but without the lid. This causes the inside wall of the cup to heat and smooth out. Both the mold and cup are cooled. The finished cup may then be removed from the cup mold.

Thermofusion is a useful process for recycling scrap polyethylene and polystyrene. Old thermoplastic products and scrap may be collected and cleaned. They can then be sorted into a polyethylene pile and a polystyrene pile. Each pile of plastic can be separately ground into a pellet form. The plastic pellets are then ready for thermofusion. The pellets may even be colored before using them for thermofusion.

REVIEW

1. Name the plastic(s) used for your thermofusion project.

2. Describe each step in the thermofusion process.

3. Define *thermofusion.*

4. Name some products made by thermofusion.

Thermofusion

5. Describe the static molding process.

6. Describe one way to recycle scrap plastic in the thermofusion process.

7. Explain why a templet is used in the thermofusion process.

8. Explain why a mold is used in the thermofusion process.

9. Name two materials used to make thermofusion molds.

10. Describe how to fix a hole or thin spot in a thermofusion product.

SELECTED BIBLIOGRAPHY

Cherry, Raymond *General Plastics Projects and Procedures.* Bloomington, Illinois: McKnight & McKnight Publishing Company, 1967.

Newman, Jay Hartley. *Plastics for the Craftsman.* New York: Crown Publishers, 1972.

Newman, Thelma R. *Crafting With Plastics.* Radnor, Pennsylvania: Chilton Book Company, 1975.

Rosato, Dominick V., ed. *Plastics Industry Safety Handbook.* Boston: Cahners Books, 1973.

EQUIPMENT AND MATERIAL SUPPLIERS

1. Ain Plastics Inc., 65 Fourth Avenue, New York, New York 10003

2. Industrial Arts Supply Company, 5724 West 36th Street, Minneapolis, Minnesota 55416.

Polystyrene may be purchased from either of the suppliers listed above. Polyethylene is available from the Industrial Arts Supply Company. Other sources include local hobby and craft stores, or recycled plastic scrap.

Assignment 9
Thermoforming

OBJECTIVES

To prepare a thermoforming mold.

To set up a thermoforming machine.

To make a vacuum formed product.

INTRODUCTION

Thermoforming is an industrial process. It is used to make a product by heating a thermoplastic sheet. The sheet is forced to conform to a mold surface or shape. The products made by thermoforming are called "thermoforms."

The thermoforming process is divided into three general types. The three types are vacuum forming, mechanical forming, and pressure forming. Because vacuum forming is more common, it is used in this assignment.

In vacuum forming, the plastic is clamped in a frame above the mold. It is heated and the mold is forced to make contact with the clamped plastic. A vacuum is then produced between the mold surface and the plastic. This causes atmospheric pressure to force the plastic against the mold. The formed part is then cooled, unclamped, and trimmed.

Fig. 9-1 shows a large, three-station industrial rotary thermoforming machine. It is being used to manufacture the house shutters shown at the right side of the picture. This is one example of an industrial thermoformed product.

SAFETY

The following precautions should be taken when thermoforming:

Fig. 9-1 Industrial rotary thermoforming machine. (Courtesy Comet International, Inc.)

Thermoforming

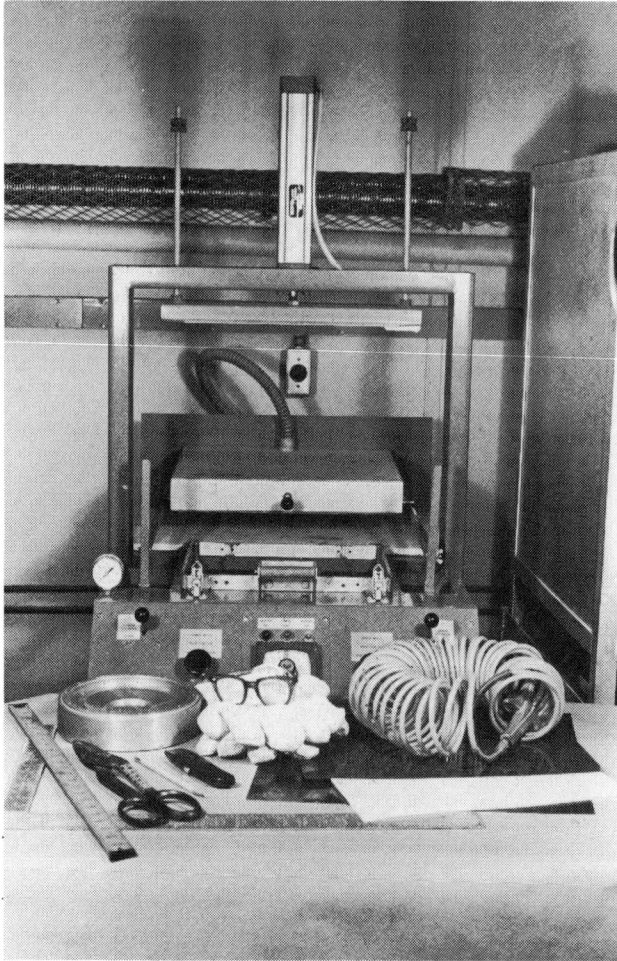

Fig. 9-2 Thermoforming equipment and materials.

Fig. 9-3 Typical thermoforming molds.

1. Work in a well-ventilated area. Do not breathe plastic fumes because some are toxic.

2. Work on a heat-resistant surface.

3. Wear safety glasses and heat-resistant gloves.

4. Keep a general purpose fire extinguisher in the work area.

5. Do not touch hot plastic or the thermoformer heater with your bare hands.

6. Learn the safe operation of the thermoforming machine.

7. Do not pinch your fingers in the thermoforming clamping mechanism.

8. Learn the safe operation of the hand and power trimming tools.

9. Do not point the high pressure air line at yourself or others.

EQUIPMENT AND MATERIALS

The equipment and materials used for thermoforming are:
1. Safety glasses.
2. Heat-resistant gloves.
3. Heat-resistant surface.
4. Thermoforming machine.
5. Mold.
6. Polyethylene or polystyrene.
7. High pressure air line to machine.
8. High pressure air line and air gun.
9. Metal or paper cutting shears.
10. Utility knife.
11. Carpenter's square.
12. Rule.
13. Pencil or marker.

THERMOFORMING LAB PROCEDURE

1. Obtain all items listed under "Equipment and Materials" needed

Section II: Molding

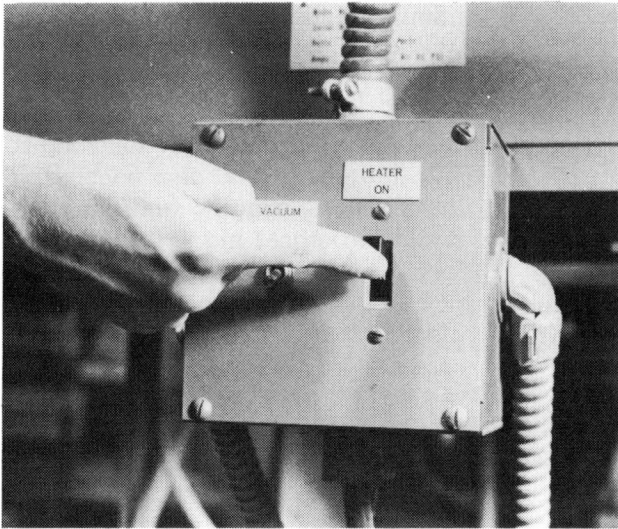

Fig. 9-4 Turning on heater switch.

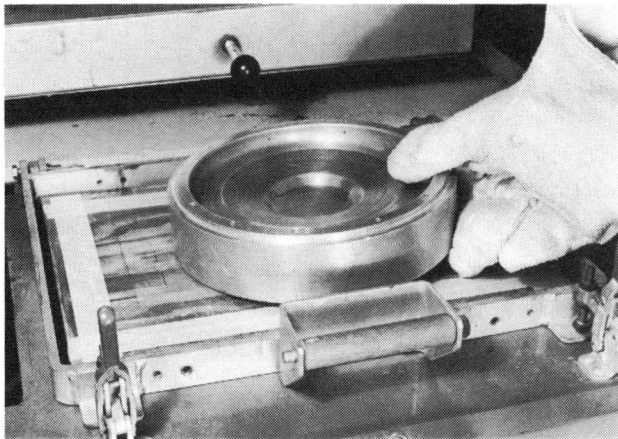

Fig. 9-5 Placing mold on platen.

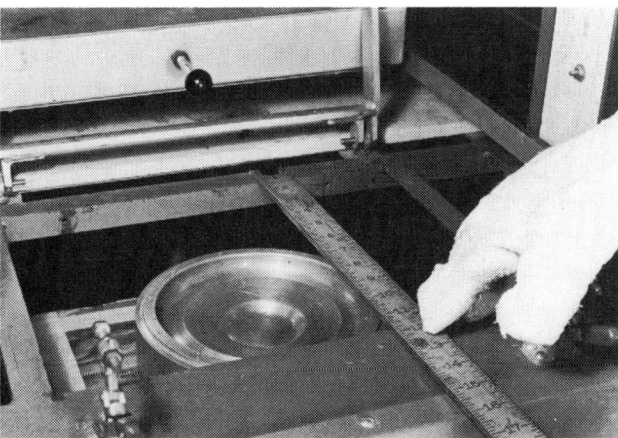

Fig. 9-6 Measuring clamping frame.

to thermoform a product. See Fig. 9-2.

2. Select the mold for this assignment. Molds are made from aluminum, close grained hard woods, epoxy steel, and epoxy aluminum. See Fig. 9-3. The flying saucer mold used in this assignment is made of aluminum.

3. Turn on the heater switch. See Fig. 9-4.

4. Wipe the inside mold surface clean.

5. Place the mold on the vacuum box or platen. See Fig. 9-5. **Make sure the mold is in the center of the platen.**

6. Measure the width and length of the clamping frame. See Fig. 9-6.

7. Lay out the width and length measurements on the plastic sheet. Use a carpenter's square, marker, and a scale.

8. Cut out the sheet along the layout lines. Use metal or paper cutting shears.

9. Place the plastic under the clamp frame. See Fig. 9-7. Note that the patterned side of the plastic is facing down. The plain side is facing you.

10. Clamp the frame tightly over the plastic.

11. Turn on the vacuum switch.

12. Wait until the vacuum gauge reads 26 or 27 inches of mercury. A higher gauge reading is also good for vacuum forming.

13. Pull the heater over the plastic. See Fig. 9-8.

14. Push the heater back every

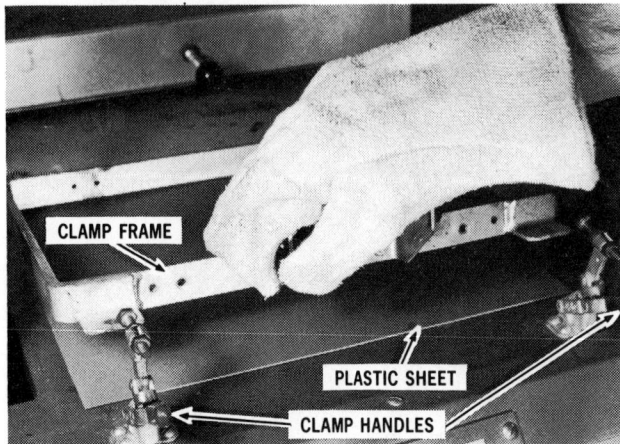

Fig. 9-7   Placing plastic under clamp frame.

Fig. 9-8   Pulling heater over plastic.

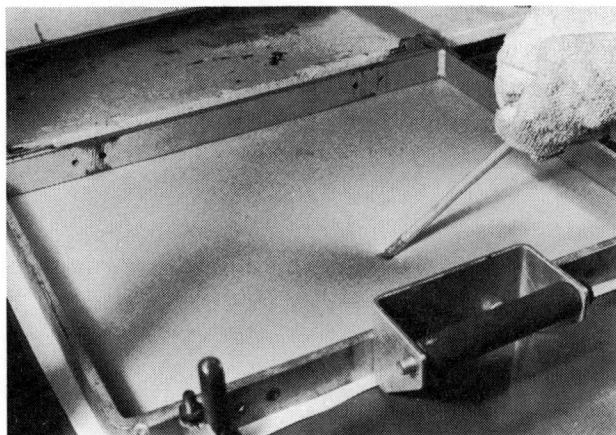

Fig. 9-9   Checking plastic.

minute to check the plastic. Some types of plastic will arch while heating.

15. Check the plastic by touching it with the eraser end of a pencil. If the plastic is sagging and does not retain the eraser mark, it is ready to form. See Fig. 9-9. **Do not heat the plastic beyond this point.** Some plastics may burn if more heat is applied.

16. Push the platen UP switch. This raises the platen and forces the mold into the hot plastic. See Fig. 9-10. Keep the mold in this position during the next step.

17. Pull the vacuum switch. This creates a vacuum between the mold surface and the plastic. It also lets the atmospheric pressure force the hot plastic against the mold surface.

18. Spray the product with cool air. Cool the product until you can touch it with your bare hand.

19. Unclamp the clamping frame.

20. Remove the mold from the plastic. Do this by hand. **Do not touch the mold surface with a tool.** This will scratch the mold.

21. Trim the extra plastic from the product. Use a utility knife. See Fig. 9-11. Metal or paper cutting shears may also be used.

Fig. 9-12 shows the completed product (a flying saucer). In Fig. 9-13 many other vacuum-formed products are shown. To make these, use a different mold and the same procedure.

CONCLUSION

Most thermoplastic sheets become rubbery after heating and rigid when cooled. This property is called *thermoelasticity*. This property

Section II: Molding

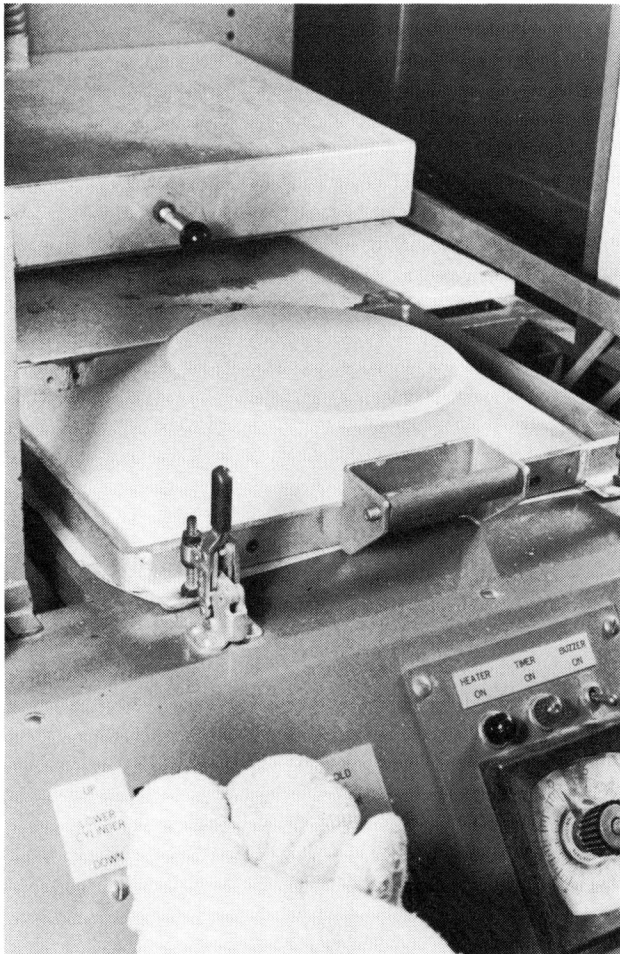

Fig. 9-10 Pushing platen switch to UP position.

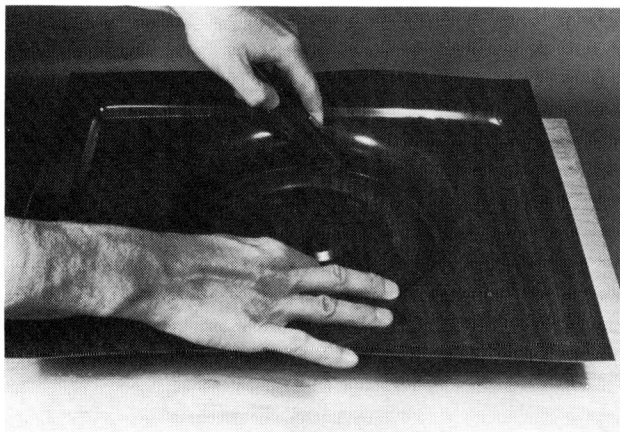

Fig. 9-11 Trimming excess plastic.

makes it possible to thermoform plastic sheets into products.

Often a complex part such as a bathroom unit for a recreational vehicle must be formed. To make such a product, vacuum, pressure, and mechanical thermoforming types are used. Each type has its advantages and limitations. The advantages of each thermoforming type require combining them to make the complex part.

Rejected or scrap thermoplastic parts can be reused. The scrap products are collected and ground into pellets in special sheet grinding machines. These pellets are mixed with new thermoplastics of the same type. They are placed into the hopper of an extruder. This machine extrudes the material into new plastic sheets. The sheets can be thermoformed.

REVIEW

1. List four products produced by the vacuum forming process.

2. Name three thermoforming processes.

3. Define a *thermoform*.

4. List three thermoforming mold materials.

5. Define *vacuum forming*.

6. Define *thermoelasticity*.

7. Name the unit of measure for vacuum.

8. Name the type of plastic used in this activity.

9. Describe how to reuse thermoplastic scrap made during the thermoforming process.

Thermoforming

Fig. 9-12 Finished product.

Fig. 9-13 Examples of vacuum-formed products.

SELECTED BIBLIOGRAPHY

Agranoff, Joan, ed. *Modern Plastics Encyclopedia.* New York: McGraw-Hill Book Company, 1976-77.

Baird, Ronald J. *Industrial Plastics.* South Holland, Illinois: The Goodheart-Willcox Company, Inc., 1971.

Milby, Robert V. *Plastics Technology.* New York: McGraw-Hill Book Company, 1973.

Patton, William J. *Plastics Technology: Theory, Design, and Manufacture.* Reston, Virginia: Reston Publishing Company, Inc., 1976.

Richardson, Terry A. *Modern Industrial Plastics.* Indianapolis: Howard W. Sams & Company, Inc., 1974.

Rosato, Dominick V., ed. *Plastics Industry Safety Handbook.* Boston: Cahners Books, 1973.

Thermoforming-Teacher's Manual. Indianapolis: Howard W. Sams & Company, Inc., 1974.

EQUIPMENT AND MATERIAL SUPPLIERS

1. AAA Plastics Equipment Company, Inc., P.O. Box 11512, Fort Worth, Texas 76110.

2. Brodhead-Garrett, 4560 East 71st Street, Cleveland, Ohio 44105.

3. Brown Plastics Engineering Company, Inc., 1823 Holste Road, Northbrook, Illinois 60062.

4. Cope Plastics, Inc., 4441 Industrial Drive, Godfrey, Illinois 62035.

Section II: Molding

5. Delvies Plastics, Inc., 2320 South West Temple, P.O. Box 1415, Salt Lake City, Utah 84110.

6. Graves-Humphreys, Inc., 1948 Franklin Road, P.O. Box 1347, Roanoke, Virginia 24033.

7. Industrial Arts Supply Company, 5724 West 36th Street, Minneapolis, Minnesota 55416.

8. McKilligan Industrial Supply Corporation, 494 Chenango Street, Binghamton, New York 13901.

9. Paxton/Patterson, 5719 West 65th Street, Chicago, Illinois 60638.

Thermoforming machines can be purchased from any of the above listed suppliers. Thermoforming molds are available from suppliers 1, 2, 5, 6, 7, 8, and 9. Plastic for thermoforming can be purchased from suppliers 2, 4, 5, 6, 7, 8, and 9.

Thermoforming

Section III
Moldmaking

Assignment 10
Silicone Molds

OBJECTIVES

To select and prepare a split pattern.

To make a two piece silicone mold.

INTRODUCTION

Silicone is an elastomer. Silicone molds can be flexed to eject parts with undercuts. Silicone is generally more expensive than other elastomers. It easily reproduces pattern detail. It also has good tear strength, excellent mold life, and moderate heat-resistance. It does not often need a release agent applied to it before being cast into the silicone mold.

Silicone molds are used to cast epoxy, plaster, polyurethane foam, polyester, concrete, candles, and low melting temperature metals. Silicone is also used for certain hot stamping dies, vacuum bags, pressure pads, electroplating masks, and sandblasting masks. Silicone molds can reproduce fine pattern detail like wood grain. The molds remain flexible enough to eject parts having some negative draft.

The silicone material used in this assignment is a two part system. The hardener is supplied in the proper proportion to the resin. This material is mixed, handled, and cured at room temperature (RTV). At room temperature, the mixed silicone cures or crosslinks into an insoluble state. Silicone can be easily used in schools.

SAFETY

The following precautions should be taken when making silicone molds:

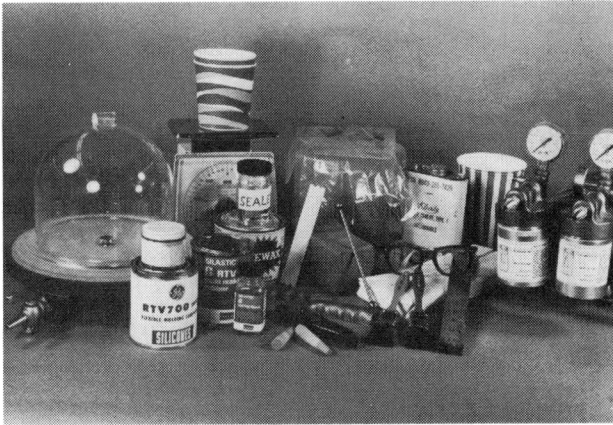

Fig. 10-1 Silicone mold making equipment and materials.

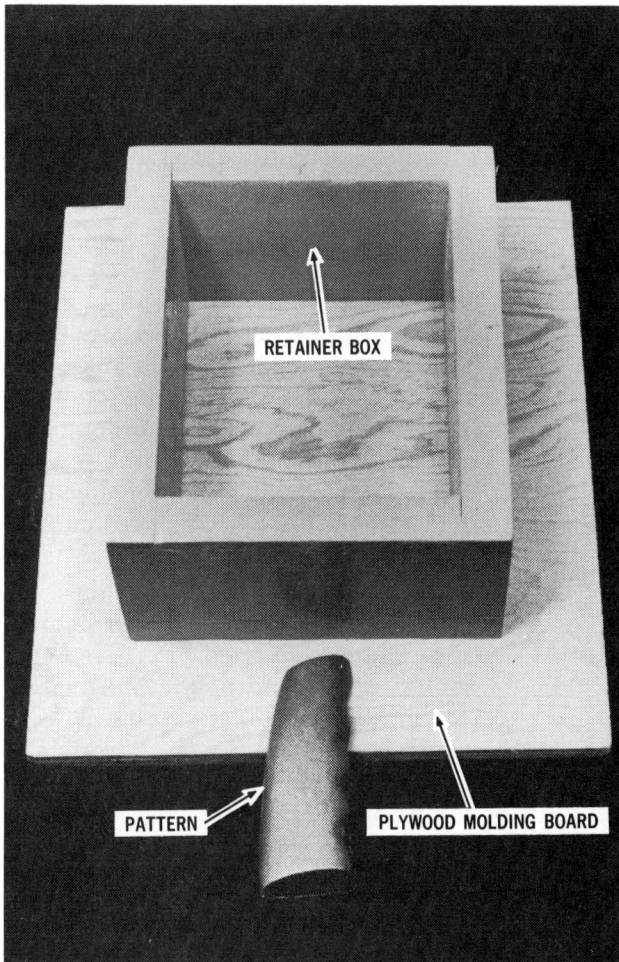

Fig. 10-2 Pattern and retainer box.

1. Work in a well-ventilated area. Do not breathe silicone vapors.

2. Wear safety glasses and disposable gloves.

3. Protect your eyes and skin from contact with silicone components. If contact is made, *immediately* wash your skin and flush your eyes with clean water for 15 minutes. Then, *get medical attention.*

EQUIPMENT AND MATERIALS

The equipment and materials needed for making silicone molds are:

1. Safety glasses.
2. Disposable (polyethylene) gloves.
3. Two-piece retainer box of wood, sheet metal, or sheet plastic.
4. Two-piece pattern.
5. Plywood molding board.
6. Silicone resin and hardener.
7. Pattern sealer.
8. Release agent (high-grade, paste-type furniture wax).
9. Rubber cement.
10. Modeling clay.
11. Fillet iron.
12. Mixing containers (plastic, glass, metal, or unwaxed paper cups).
13. Wood stirring stick or spatula.
14. Vacuum degassing equipment.
15. Measuring rule.
16. Weight scale.
17. 2 one-inch brushes.
18. Cloth or paper towels.

SILICONE MOLD LAB PROCEDURE

1. Obtain all the items listed under "Equipment and Materials" needed to make a silicone mold. See Fig. 10-1.

2. Obtain the pattern and retainer box. See Fig. 10-2. The pattern should have two holes in its back

Fig. 10-3   Placing pattern in retainer box.

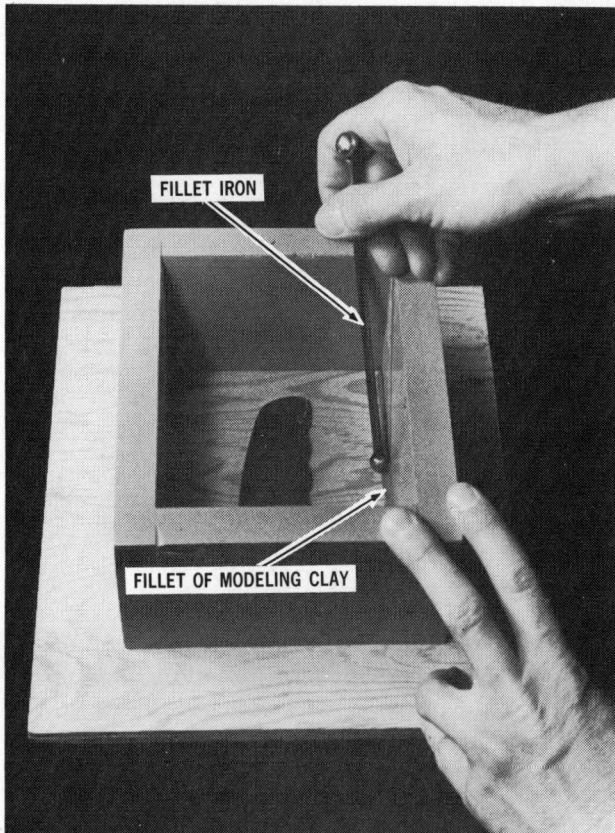

FILLET IRON

FILLET OF MODELING CLAY

Fig. 10-4   Placing fillet of modeling clay.

side. The retainer box should have two holes in its edge. Patterns are made of wood, glass, glazed ceramic, metal, high density plaster and some plastics. Retainer boxes are made of wood, sheet metal, or sheet plastic. Make sure the retainer box is fastened to the molding board with the two-hole edge down.

3. Place rubber cement on the pattern back. This stops the pattern from floating when silicone is cast around it.

4. Place the pattern in the retainer box. See Fig. 10-3. The distance of the pattern from the retainer box determines the mold wall thickness. This distance should be ½". Make sure the pattern lies flat on the molding board.

5. Place a fillet of clay around all inside retainer box corners. See Fig. 10-4. Make the fillet with a fillet iron. The fillets stop the cast silicone from leaking from the retainer. They also cause a smooth outside corner to be cast in the silicone.

6. Brush a bubble-free coat of a manufacturer's brand sealer on the pattern, retainer box, fillets, and molding board. This is done only if these items contain large amounts of rosins, tars, sulfur containing clays, organic adhesives, or finishes that prevent the silicone from curing.

7. Apply several coats of release agent to the pattern, molding board, and retainer box. See Fig. 10-5. Let each coat dry and gently buff before applying the next coat. Some RTV silicones require no release agents. Your instructor will tell you when to use a release agent.

8. Calculate the volume of the retainer box. See Fig. 10-6. This is done by measuring its width, depth, and length, and then multiplying them together. Estimate the volume

Section III: Moldmaking

to be filled with silicone. This is done by subtracting the pattern volume from the retainer volume.

9. Check the silicone manufacturer's weight-to-volume specifications. This specification will differ for each silicone. A manufacturer's specifications may state that a pound of silicone will fill so many cubic inches of space.

10. Weigh the amount of resin and hardener. See Fig. 10-7. Ask your instructor how much hardener to use. The amount of hardener will differ for each manufacturer's silicone. For example, some silicone components are mixed 20 to 1, 10 to 1, or 5 to 1 by weight. The mixing container should be four times the volume of the mixed silicone. This lets the silicone rise without spilling during degassing.

11. Pour the hardener into the container of silicone. See Fig. 10-8. Use disposable gloves when handling silicone parts.

12. Mix the two components well until all the streaks disappear. See Fig. 10-9. **Try not to trap air in the material.** Be sure to scrape the sides and bottom of the container with the stirrer. This helps completely mix the silicone parts. Ask your instructor how long you must mix the parts before they start to harden. This is called their "pot life".

13. Degas silicone with a vacuum degassing unit. See Fig. 10-10. Make sure the container is about four times larger than the original mixed silicone. The escaping air will make the silicone rise. Degassing is finished in about two minutes or when the foaming stops. Use about 28 to 29 inches of mercury of vacuum.

14. Brush a bubble-free coat of silicone on the pattern. See Fig.

Fig. 10-5 Applying release agent to pattern.

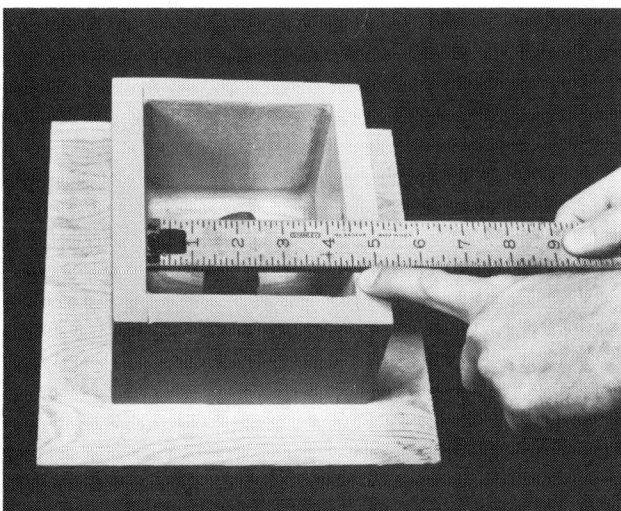

Fig. 10-6 Calculating volume of retainer box.

Silicone Molds

Fig. 10-7 Weighing out resin and hardener.

Fig. 10-8 Pouring hardener into silicone.

10-11. This helps remove air bubbles from the pattern surface. Use this method if the silicone was not vacuum degassed.

15. Pour the silicone into a corner of the retainer box. See Fig. 10-12. Let it slowly flow over the pattern. The silicone should be poured until it is about 1″ over the highest part of the pattern. If the retainer is small, it can be filled with silicone. The pouring method shown helps stop air from being trapped against the pattern surface.

16. Let the silicone cure. Ask your instructor how long the cure time should be. Cure methods are different for each type of silicone. One type of silicone cures in 24 hours at room temperature. Other silicones must be heated to cure.

17. Turn the retainer box up-side-down with the molding board removed.

18. Cut a locating dimple on each side of the mold. See Fig. 10-13. Use a utility knife.

19. Put the doweled half of the pattern on the undoweled half.

20. Put the retainer box and two tapered dowels in place. See Fig. 10-14. The dowels form cast-in-place pouring channels. Make sure each dowel is placed at the end of a runner location. Also, make sure the tapered end is toward the pattern.

21. Put fillets and a sealer on all parts as shown in Steps 5 and 6.

22. Apply a release agent to the retainer, pattern, dowels, and cast silicone mold. This was shown in Step 7.

23. Repeat Steps 8 through 16. See Fig. 10-15.

Section III: Moldmaking

Fig. 10-9 Mixing hardener and silicone.

Fig. 10-10 Degassing silicone.

Fig. 10-11 Applying silicone to pattern.

Fig. 10-12 Pouring silicone into retainer box.

Silicone Molds

Fig. 10-13   Cutting dimple.

Fig. 10-14   Adding retainer box and dowels.

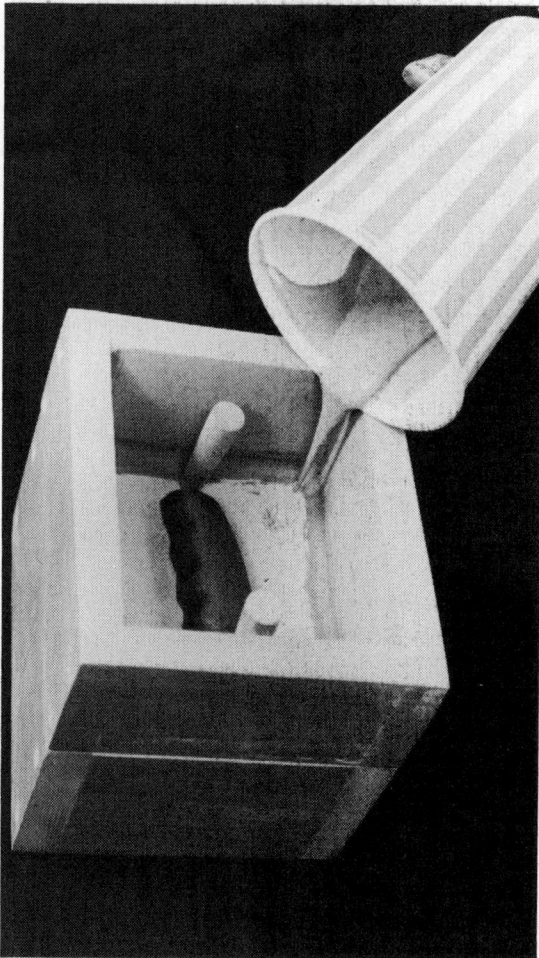

Fig. 10-15   Pouring silicone.

Fig. 10-16   Pulling molds from retainer box.

Section III: Moldmaking

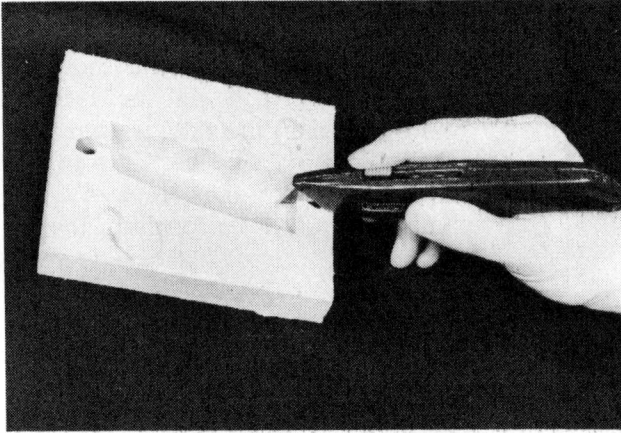

Fig. 10-17 Cutting runners (channels).

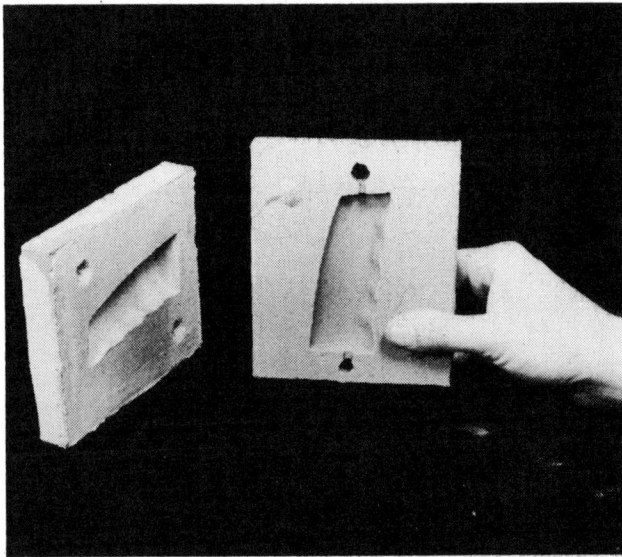

Fig. 10-18 Checking finished mold.

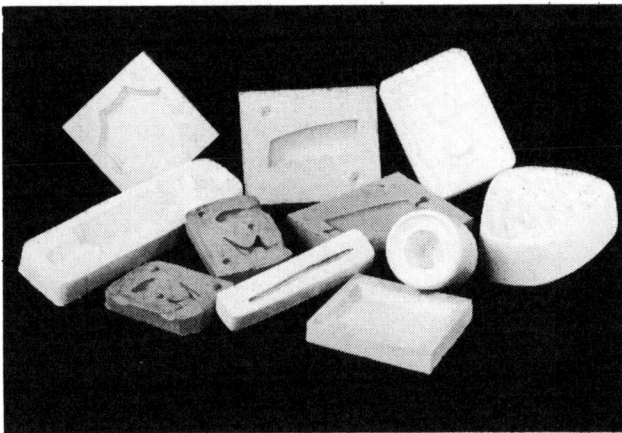

Fig. 10-19 Examples of silicone molds.

24. Separate the retainer box halves.

25. Pull the two dowels (sprue and riser) from one silicone mold.

26. Pull each mold from the retainer box. See Fig. 10-16. The retainer boxes may have to be unscrewed to remove the silicone molds. **Do not pry the silicone molds from the retainer boxes with sharp tools.**

27. Lift or peel each mold from the pattern. Use a slow or continuous pulling action. **Do not jerk the mold from the pattern.**

28. Cut two runners (channels) in the undoweled mold. See Fig. 10-17. A runner is cut from each doweled mold opening to the nearest mold cavity.

29. Check the finished mold. See Fig. 10-18. If the mold has soft or uncured spots, the silicone components were not well mixed. The silicone could have also reacted with chemicals in the pattern. This mold is ready to use. Polyester can be poured into a sprue and through a runner into the mold cavity. This will make a cast gear shift handle. The mold can also be used to urethane foam this product.

Several silicone molds are shown in Fig. 10-19.

CONCLUSION

Many silicone mold making materials are the RTV type. Some silicones require a high temperature cure cycle. Each silicone system has its advantages. Selecting the proper silicone requires a knowledge of the silicone tear strength, flexibility, hardness, and cure reaction with the pattern.

When a large silicone mold is to be built, special considerations must be

Silicone Molds

made. Cured waste or old silicone molds can be cut into small parts. They can be used as a bulking agent in large molds. This material is mixed with new silicone. It is then placed in the retainer box over partially cured silicone. Care must be taken to stop the lumps of bulking agent from touching the pattern.

Left-over mixed silicone can often be frozen. It can then be re-used the next day. This lets the moldmaker reduce the cost of his silicone.

REVIEW

1. Name some uses for silicone molds.

2. Describe a silicone elastomer mold.

3. Name two silicone mold pattern materials.

4. Describe each step in silicone mold making.

5. List the steps in preparing a pattern for silicone mold making.

6. Name two materials that may be cast and cured in silicone molds.

7. Explain how the volume-to-weight ratio and the resin-to-hardener ratio is calculated for your silicone mold.

8. Name two advantages of silicone molds.

9. Explain the purpose of a sealer coat.

10. Describe how a silicone mold bulking agent is made.

Section III: Moldmaking

SELECTED BIBLIOGRAPHY

Baird, Ronald J. *Industrial Plastics.* South Holland, Illinois: The Goodheart-Willcox Company, Inc., 1971.

Dow Corning Corporation. *Application Ideas Using Silicone Moldmaking Materials.* Bulletin No. 61-244. Midland, Michigan: Dow Corning Corporation, 1974.

_____. *Tooling with Silastic® and Dow Corning® RTV Silicone Rubber.* Bulletin No. 61-223. Midland, Michigan: Dow Corning Corporation, 1973.

Flexible Mold Making—Teacher's Manual. Indianapolis: Howard W. Sams & Company, Inc., 1974.

General Electric Company. *The Moldmakers.* Bulletin No. S-45. Waterford, New York: General Electric Company, n.d.

Milby, Robert V. *Plastics Technology.* New York: McGraw-Hill Book Company, 1973.

Morse, George T. *How to Make Flexible Molds.* Gillette, New Jersey: Smooth-On, Inc., 1974.

Patton, William J. *Plastics Technology: Theory, Design, and Manufacture.* Reston, Virginia: Reston Publishing Company, Inc., 1976.

Richardson, Terry A. *Modern Industrial Plastics.* Indianapolis: Howard W. Sams & Company, Inc., 1974.

Rosato, Dominick V., ed. *Plastics Industry Safety Handbook.* Boston: Cahners Books, 1973.

Silicone Molds

EQUIPMENT AND MATERIAL SUPPLIERS

1. Ain Plastics, Inc., 65 Fourth Avenue, New York, New York 10003.

2. Delvies Plastics, Inc., 2320 South West Temple, P.O. Box 1415, Salt Lake City, Utah 84110.

3. Dow Chemical Company, 2020 Dow Center, Midland, Michigan 48640.

4. General Electric Company, Silicone Products Department, Waterford Mechanicville Road, Waterford, New York 12188.

Silicone mold making equipment and materials can be purchased from any of the suppliers listed above.

Assignment 11
Urethane Molds

OBJECTIVES

To select and prepare a pattern.

To make a urethane mold.

INTRODUCTION

A urethane mold is a cavity. Materials are placed into this cavity to cure and form a completed product. The mold is an *elastomer*, so it is very flexible. It can be flexed to remove parts with undercuts. Urethane is often used to make large molds because the mold detail will be excellent and the tear strength good. Urethane is not as expensive as silicone rubbers.

Urethane is used for casting molds for cement-like materials. Examples are 4' × 10' precast concrete panels. The molds reproduce fine pattern detail such as wood grain. Thus, urethane can be used as a casting resin and foam casting mold material.

The urethane used in this assignment is a two part system. The hardener is supplied in the proper proportion to the resin. This material is mixed, handled, and cured at room temperature (RTV). At room temperature, the mixed urethane cures or crosslinks into an insoluble state. Urethane molds have a relatively long service life.

SAFETY

The following precautions should be taken when making urethane molds:

1. Work in a well-ventilated area. Do not breathe urethane fumes.

2. Wear safety glasses and disposable gloves.

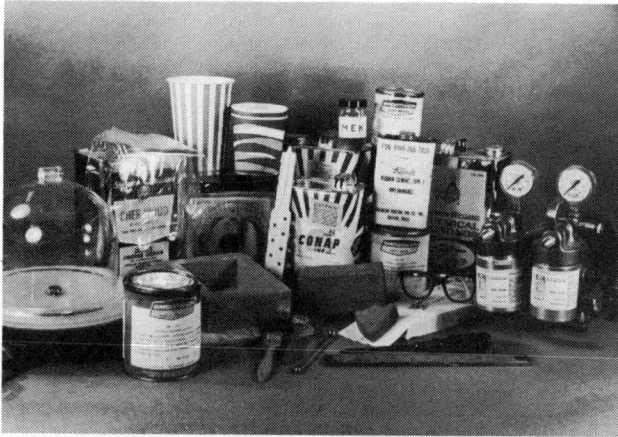

Fig. 11-1 Urethane mold making equipment and materials.

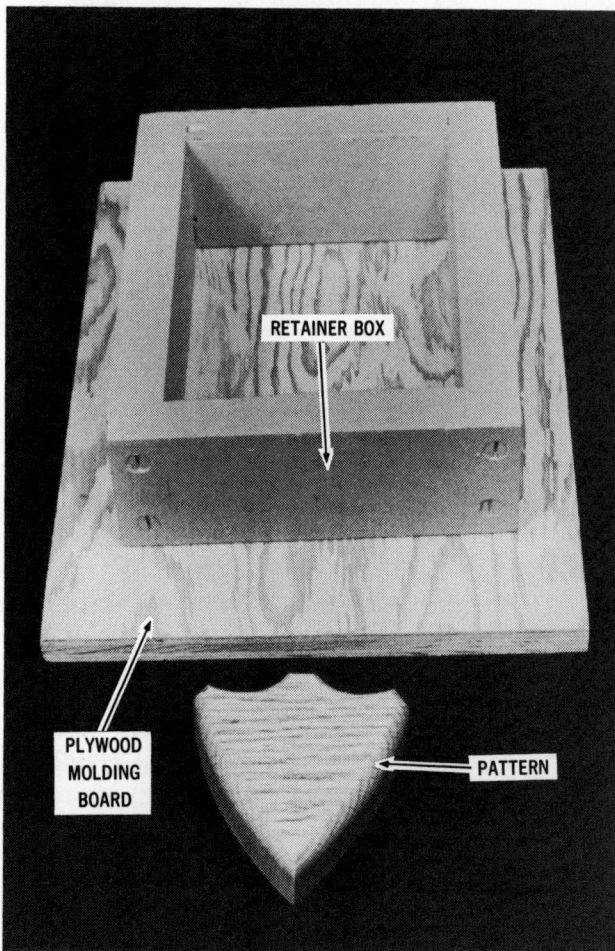

Fig. 11-2 Pattern and retainer box.

3. Protect your eyes and skin from contact with urethane components. If contact is made, *immediately* wash your skin and flush your eyes with clean water for 15 minutes. Then, *get medical attention.*

EQUIPMENT AND MATERIALS

The equipment and materials needed for making urethane molds are:

1. Safety glasses.
2. Disposable (polyethylene) gloves.
3. Retainer box (wood, sheet metal, linoleum, or heavy cardboard).
4. Pattern.
5. Molding board (plywood).
6. Urethane and hardener.
7. Release agent.
8. Rubber cement or flathead woodworking screws.
9. Modeling clay.
10. Fillet iron.
11. Mixing containers (glass, metal, or polyethylene).
12. Stirring rods (metal, plastic, or glass).
13. Vacuum degassing equipment.
14. Measuring rule.
15. Weight scale.
16. 2 one-inch brushes.
17. Urethane solvent or MEK.
18. Lacquer and container.
19. Lacquer thinner.
20. Screwdriver.
21. Cloth or paper towels.

URETHANE MOLD LAB PROCEDURE

1. Obtain all the items listed under "Equipment and Materials" needed to make a urethane mold. See Fig. 11-1.

2. Obtain the pattern and retainer box. See Fig. 11-2. Patterns are made of wood, glass, glazed ceramic, soap, metal, high-density

Section III: Moldmaking

Fig. 11-3 Centering pattern in retainer box.

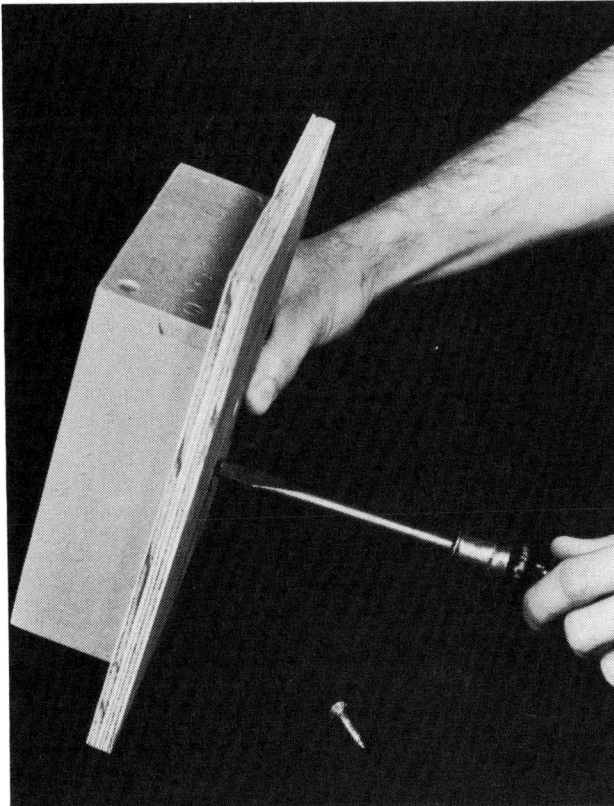

Fig. 11-4 Fastening pattern to molding board.

plaster, polyester, or epoxy. Retainer boxes are made of wood, sheet metal, linoleum, or heavy cardboard. Make sure the retainer box is taped, rubber cemented, nailed, or screwed to the molding board.

3. Center the pattern in the retainer box. See Fig. 11-3. The distance of the pattern from the retainer box determines the wall thickness of the mold. This distance should be ½". Make sure the pattern lies flat on the molding board.

4. Fasten the pattern to the molding board with rubber cement or woodworking screws. See Fig. 11-4. This stops the pattern from floating when urethane is cast around it.

5. Place a fillet of modeling clay around the pattern bottom and on all inside retainer box corners. See Fig. 11-5. Make the fillet with a fillet iron. The fillets stop the cast urethane from running under the pattern and forcing out air bubbles. They also cause a smooth outside corner to be cast in the urethane.

6. Heat the pattern, molding board, and retainer box to 90-110° F. This is an extra step. It is only done if the pattern, retainer, and molding board are porous. The wood being used here is porous. These parts may contain water and, so, they must be dried.

7. Brush acrylic lacquer on the pattern, retainer box, fillets, and molding board. This stops the urethane from reacting with oils, paints, varnishes, water, and old release agents on these parts.

8. Apply several coats of release agent to the pattern, molding board, and retainer box. Let each coat dry and gently buff before applying the next coat.

Urethane Molds

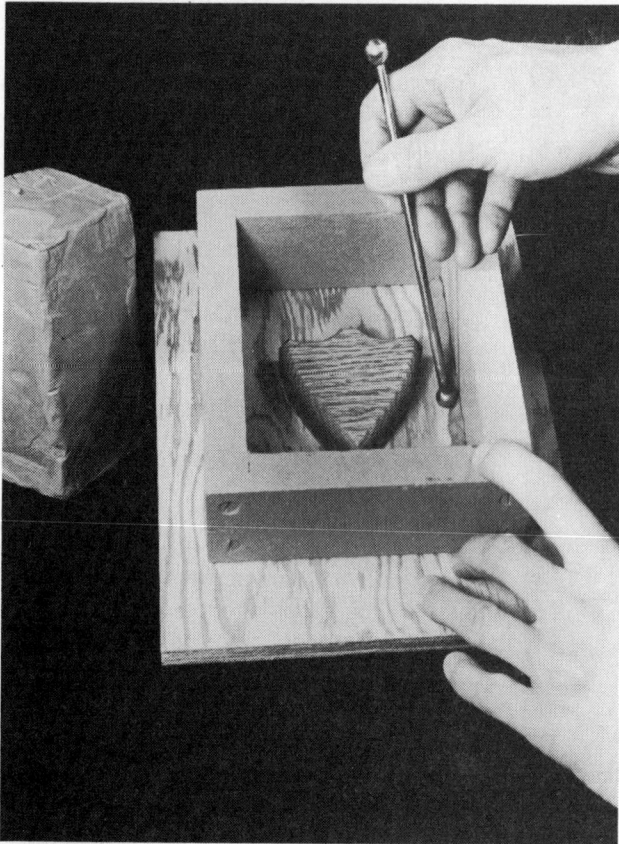

Fig. 11-5 Making fillet with fillet iron.

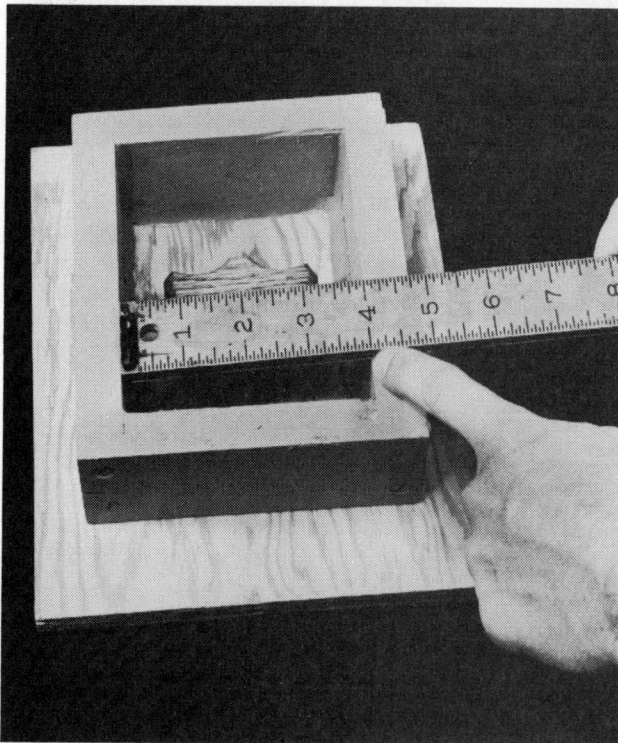

Fig. 11-6 Calculating volume of retainer box.

9. Determine the volume of the retainer box. See Fig. 11-6. This is done by first measuring the width, depth, and length of the box. Then, multiply these together to find the volume. Estimate the volume of the pattern and subtract it from the retainer volume figure. This answer is the volume to be filled with urethane.

10. Fill the retainer box with water. Now, pour the retainer box water into a volume calibrated (marked) container. This tells you the size of the space (volume) taken up by the retainer box. This is another way to figure the volume of the retainer box. Be sure to dry the retainer box before going on to the next step.

11. Check the urethane manufacturer's weight-to-volume ratio specifications. These specifications will differ for each type of urethane. A manufacturer's specifications may say that a pound of urethane will fill so many cubic inches of space.

12. Weigh the amount of resin and hardener needed. See Fig. 11-7. Your instructor will tell you how much hardener to use. The amount of hardener will differ for each manufacturer's urethane. Some urethane parts are mixed one to one by weight. Others are mixed 100 parts to 50 parts by weight.

13. Pour the hardener into the container of resin, as in Fig. 11-8. Use an oversized, unwaxed paper, metal, glass, or plastic container. A large container will let you vacuum degas the urethane. Use disposable gloves when handling urethane parts.

14. Mix the two components well. See Fig. 11-9. Try not to trap air in the material. Be sure to scrape the sides and bottom of the container with the stirrer. This helps

Section III: Moldmaking

Fig. 11-7   Weighing out resin and hardener.

Fig. 11-8   Pouring hardener into resin.

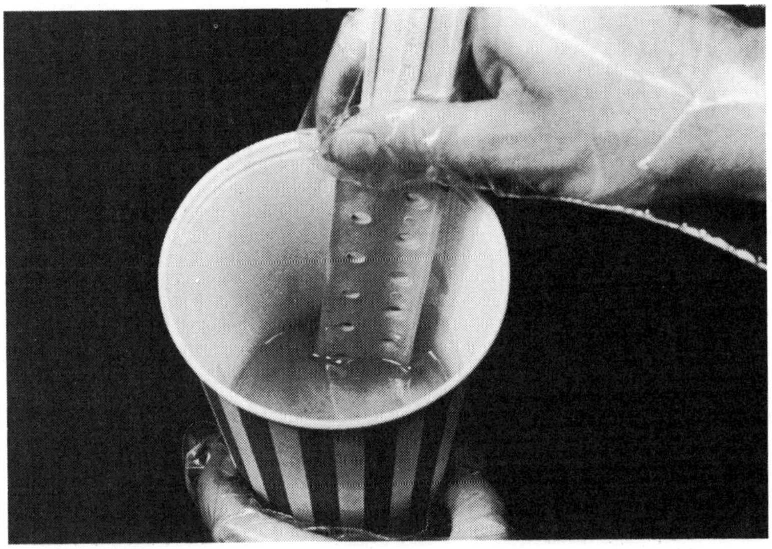

Fig. 11-9   Mixing hardener and resin.

Urethane Molds

completely mix the urethane parts. Your instructor will tell you how long to mix the parts before they start to harden (pot life).

15. Degas the urethane with a vacuum degassing unit. See Fig. 11-10. Make sure the container is about four times larger than the mixed urethane. The escaping air will make the urethane rise. Degassing is finished in about two minutes or when the foaming stops. Use about 28-29 inches of mercury of vacuum.

16. Brush a bubble-free coat of urethane on the pattern. See Fig. 11-11. This helps remove air bubbles from the pattern surface. Use this procedure if the urethane was not degassed. Be sure to clean the brush in MEK immediately.

17. Pour the urethane into a corner of the retainer box. See Fig. 11-12. Let it slowly flow over the pattern. The urethane should be poured until it is about 1 inch over the highest part of the pattern. If the retainer is small, it can be filled with urethane. The pouring method shown helps stop air from being trapped against the pattern surface.

18. Let the urethane cure. The instructor will tell you the cure time. Cure techniques are different for each type of urethane. One type of urethane cures for one hour at 77° F. Then it has to be heated to 176° F for 16 hours. It can also be cured in 7 days at 77° F. Sometimes urethanes can be removed from the retainer before they are cured. One type of urethane can be demolded after 2 hours at 176° F or 4 hours at 77° F.

19. Loosen the urethane with finger pressure from the corners of the retainer. See Fig. 11-13.

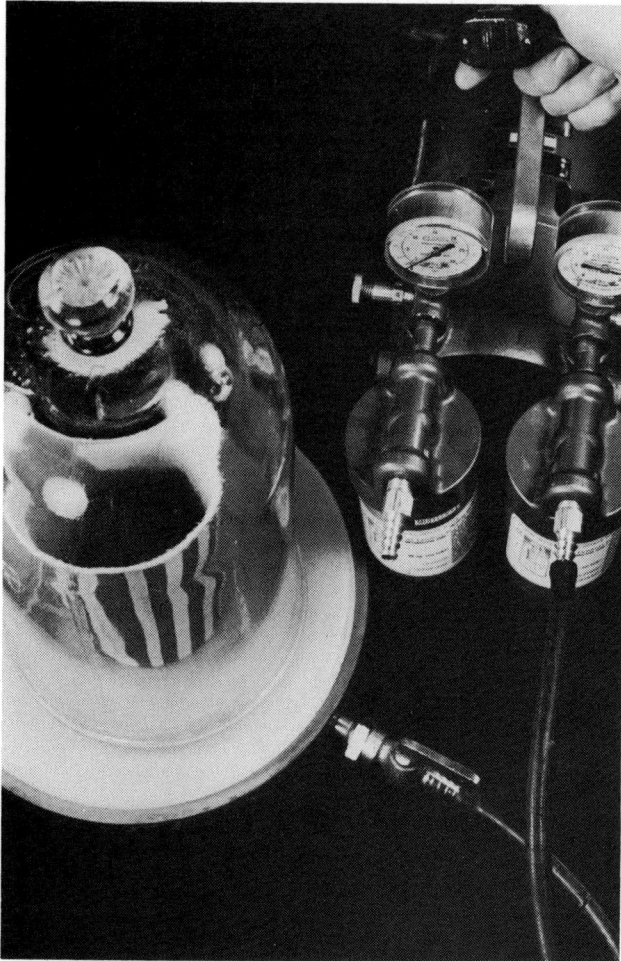

Fig. 11-10 Degassing the urethane.

Section III: Moldmaking

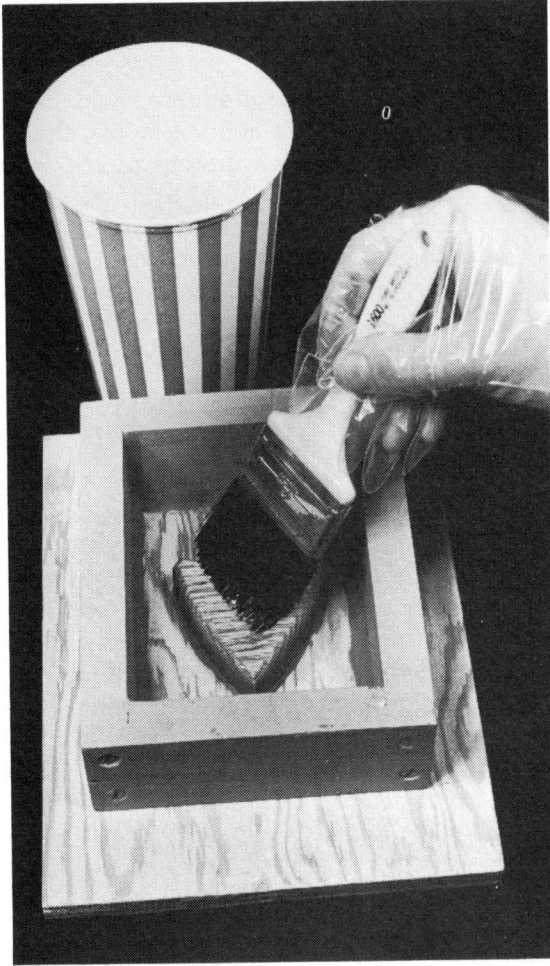

Fig. 11-11 Applying coat of urethane to pattern.

Fig. 11-12 Pouring urethane into retainer box.

Fig. 11-13 Loosening urethane with fingers.

Urethane Molds

20. Peel or flex the urethane from the pattern.

21. Check the urethane mold. See Fig. 11-14. If it has soft spots, the urethane parts were not mixed well.

22. Remove the release agent from the mold. A strong detergent or soap may be used. The mold is ready for use. Be sure to place a base coat and a release agent on the mold. This should be done if it is to be used for a polyester cast or urethane foamed product.

Shown in Fig. 11-15 are several urethane molds and their patterns. Some urethane molds are light in color and others are dark. Two-piece urethane molds can be made as shown in the Silicone Moldmaking assignment. Two piece urethane molds can be made to cast furniture parts.

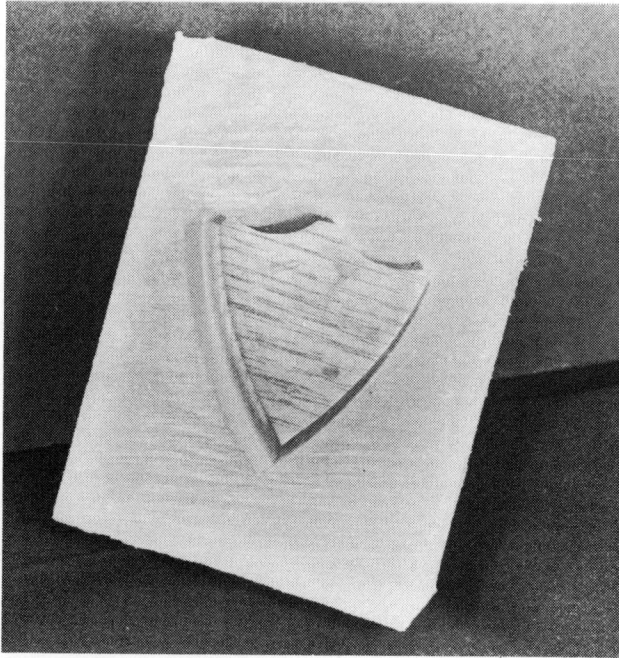

Fig. 11-14 Checking urethane mold.

CONCLUSION

Many urethane mold making materials are the RTV type. Some urethanes require high temperature curing cycles over long periods of time. Each urethane has its advantages.

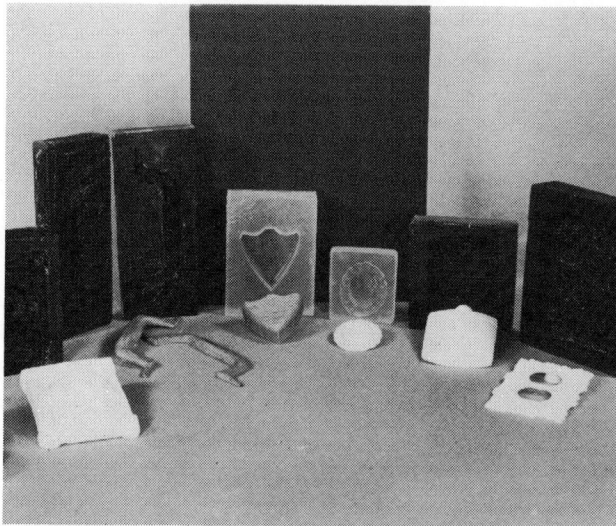

Fig. 11-15 Examples of urethane molds and patterns.

Urethane molds used to cast furniture parts must be made and used as follows. A special mold release should be put on the pattern before casting the urethane mold. The release agent is designed so it will not chemically react with the completed urethane mold. Also, the release agent will not chemically react with the furniture part surface cast into it. The completed urethane mold cavity should be cleaned of all mold release. It should be coated with a base coat. The base lets the cast furniture part receive a furniture finish. The urethane manufacturer can supply the information needed to purchase and use the special release agents and base coats.

Section III: Moldmaking

The urethane elastomer mold making procedure is economical. Urethane molds are more expensive than latex, but they last longer. They also reproduce much finer detail. When compared to silicone rubber making materials, urethane is generally cheaper.

REVIEW

1. Name two uses for urethane molds.

2. Describe a urethane elastomer mold.

3. Name two urethane mold pattern materials.

4. Describe each step in urethane mold making.

5. List the steps in preparing a pattern for urethane mold making.

6. Name two plastics that can be cast and cured in urethane molds.

7. Describe how you calculated the volume to weight ratio, and the resin to hardener ratio for your urethane mold.

8. Name two advantages of urethane molds.

9. Describe two urethane mold curing processes.

10. Explain the purpose of a base coat.

11. List two materials of which urethane mold retainer boxes are made.

12. Describe the procedure for sealing a painted, porous, wooden pattern.

Urethane Molds

SELECTED BIBLIOGRAPHY

Conap, Inc. *Conap.* Bulletin TU-80, Olean. New York: Conap, Inc.,

Devcon Corporation. *Devcon, Flexane.* Bulletin D-13. Danvers, Massachusetts: Devcon Corporation, 1975.

Flexible Mold Making—Teacher's Manual. Indianapolis: Howard W. Sams & Company, Inc., 1974.

Milby, Robert V. *Plastics Technology.* New York: McGraw-Hill Book Company, 1973.

Morse, George T. *How to Make Flexible Molds.* Gillette, New Jersey: Smooth-On, Inc., 1974.

Rosato, Dominick V., ed. *Plastics Industry Safety Handbook.* Boston: Cahners Books, 1973.

EQUIPMENT AND MATERIAL SUPPLIERS

1. Conap, Inc., 1405 Buffalo Street, Olean, New York 14760.

2. Cope Plastics, Inc., 4441 Industrial Drive, Godfrey, Illinois 62035.

3. Industrial Arts Supply Company, 5724 West 36th Street, Minneapolis, Minnesota 55416.

4. Smooth-On, Inc., 1000 Valley Road, Gillette, New Jersey 07933.

5. The Dexter Corporation, Hysol Division, Olean, New York, 14760.

Urethane mold materials may be purchased from any of the suppliers listed above.

Assignment 12
Latex Molds

OBJECTIVES

To select and prepare a flat back pattern.

To make a latex mold.

INTRODUCTION

A latex mold is easy to bend because it is made from rubber. It is used to hold cast plastics or plasters. A latex mold can be stretched to remove plastic or plaster parts.

The latex molding process works as follows. Select a pattern that will make a good latex mold. The pattern may be of ceramic, cement, plaster, wood, plastic, or metal. Do not use brass or metals containing copper. These metals destroy the flexibility and strength of latex molds. The patterns may be one or two pieces. They should have a good draft. Good draft prevents the latex mold from tearing when it is pulled from the pattern. After selecting a good pattern, it must be coated with mold release.

The pattern is dipped, sprayed, or brushed with a latex coat. Each latex coat must cure before the next one is applied. Fifteen to thirty coats of latex may be placed on the pattern before the mold is strong enough. Reinforcing cloth may be added after the eleventh latex coat is applied. This gives the mold strength. The reinforcing material is plastic mesh, gauze, cheese cloth, or burlap. After the last latex layer has cured, the mold is peeled from the pattern. A release agent is then put on the latex mold. Plaster or plastic can now be cast into the latex mold.

SAFETY

The following precautions should be taken when making latex molds:

1. Work in a well-ventilated area. Do not breathe latex fumes.

2. Wear safety glasses and disposable gloves.

3. Protect your eyes and skin from contact with latex. If contact is made, immediately wash your skin and flush your eyes with clean water for 15 minutes. Then, get medical attention.

EQUIPMENT AND MATERIALS

The equipment and materials for making latex molds are:

1. Safety glasses.
2. Disposable gloves.
3. Retainer box for plaster.
4. Pattern.
5. Latex.
6. Reinforcing material (gauze or cheese cloth).
7. Mold release agent (furniture wax).
8. High-density plaster.
9. Plaster release agent (antiseptic liquid green soap).
10. Container of soapy water solution.
11. Shellac.
12. 2 brushes (½" and 1" sizes).
13. Scissors.
14. Unwaxed paper cups.
15. Clean cloth or paper towels.

Fig. 12-1 Latex mold making equipment and materials.

Fig. 12-2 Latex mold pattern.

LATEX MOLD LAB PROCEDURE

1. Obtain all the items listed under "Equipment and Materials" needed to make a latex mold. See Fig. 12-1.

2. Obtain the pattern to be coated with latex. See Fig. 12-2. The pattern

Fig. 12-3 Applying a coat of shellac.

Fig. 12-4 Applying a latex coat.

Fig. 12-5 Applying more latex.

used in this assignment is made of high density plaster.

3. Place a coating of shellac or lacquer on the plaster pattern. See Fig. 12-3. Let the coating thoroughly dry. The shellac or lacquer seals the pattern pores. This stops the latex from soaking into the plaster.

4. Place a coat of release agent (furniture wax) on the pattern. Let the release agent dry and lightly buff it.

5. Brush a smooth, even coat of latex on the pattern. See Fig. 12-4. Be sure to brush out all air bubbles. Let each latex coat dry (cure) before applying the next coat. Ask your instructor how long drying should take. It takes a few days to place all the latex coats on the pattern. The more latex coats placed on the pattern, the stronger the mold will be. Be sure to clean the brush in soapy water after each latex coating.

6. Brush on more coats of latex. See Fig. 12-5. Follow the procedure shown in Step 5. Be sure each coat cures before applying the next coat. It may take 20 to 30 latex coats to make a strong mold.

7. Brush a coat of latex through a reinforcing material. See Fig. 12-6. The material (gauze or cheesecloth) is applied after the eleventh layer of latex is applied. Several layers of reinforcing material may have to be applied to make the mold strong.

8. Place the final latex coat on the pattern as described in Step 5. See Fig. 12-7.

9. Cure the latex mold on the pattern at room temperature for three days.

10. Press your fingernail into the latex. If the mark or depression snaps back, the latex is cured.

Latex Molds

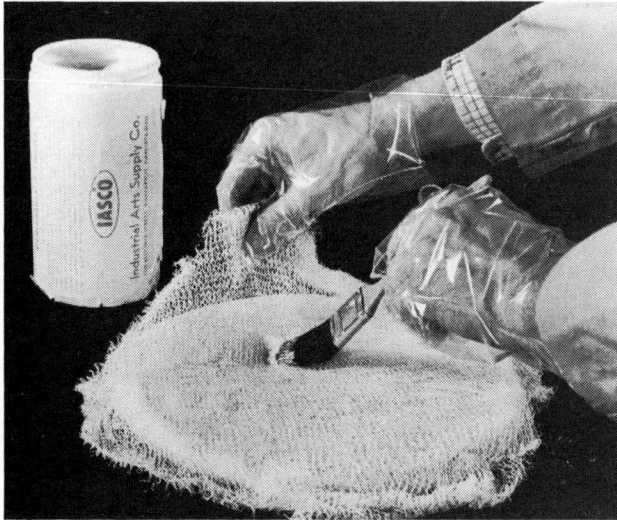

Fig. 12-6 Brushing latex through reinforcing material.

Fig. 12-7 Applying the final latex coat.

11. Place a retainer of wood, heavy cardboard, or plastic around the latex coated pattern.

12. Brush a plaster release agent on all exposed surfaces. See Fig. 12-8. Antiseptic green soap is the plaster release agent used in this lab exercise.

13. Mix plaster according to your instructor's directions.

14. Pour the plaster into the retainer and around the latex. See Fig. 12-9.

15. Let the plaster cure. Other latex mold back-up materials can be used. The plaster mold back-up adds strength to the latex mold. It also helps the latex mold hold its shape when it is cast full of material.

16. Turn the unit (retainer, plaster back-up, pattern, and mold) over.

17. Remove the latex mold and pattern together from the plaster back-up. See Fig. 12-10. Hold the pattern and latex mold together during this step.

18. Peel or flex the latex mold from the pattern. See Fig. 12-11. Be careful not to tear the latex. Latex molds are easily torn.

19. The latex mold is now complete. See Fig. 12-12. Be sure to place the mold in its back-up before using it. Coat the mold with your instructor's recommended release agent. This must be done before casting polyesters or urethane resins in it. These resins can damage an uncoated latex mold.

CONCLUSION

The latex molding process is very inexpensive. Rubber latex elastomer molds will provide a long, useful

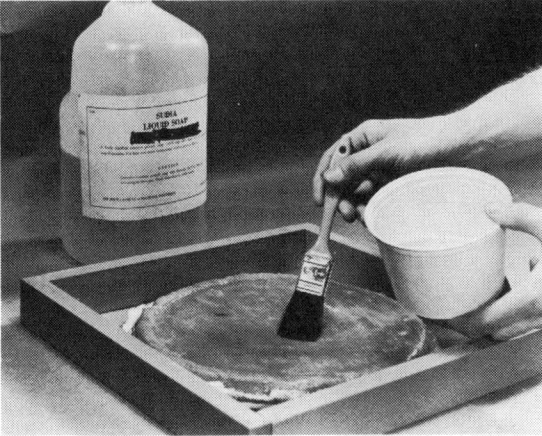

Fig. 12-8   Applying plaster release agent.

Fig. 12-9   Plaster poured into retainer.

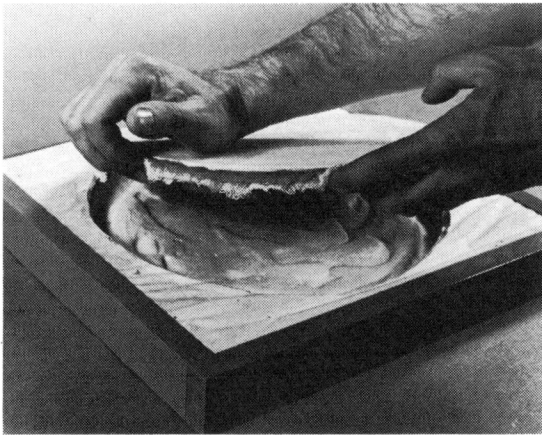

Fig. 12-10   Removing latex mold and pattern.

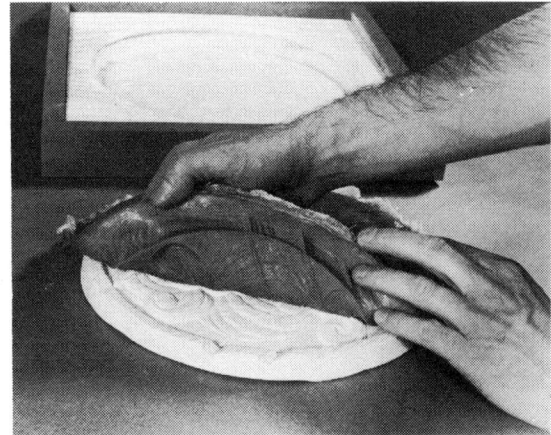

Fig. 12-11   Peeling mold from pattern.

Fig. 12-12   Completed latex mold.

Fig. 12-13   Examples of latex molds.

Latex Molds

life. They are less expensive to make than silicone and urethane molds or matched metal molds. Many latex molds are shown in Fig. 12-13. Use the procedure just described to make them.

Rubber latex elastomer mold materials are room temperature vulcanizing (RTV). This means that they will crosslink (set up) at room temperature. They also form into a nondissolving and nonshrinking material.

Latex is used when a large mold is to be made. It is also used when mold detail is to be of average reproduction quality. Models, statuettes, and carvings with undercuts can be made with latex molds. Latex molds are inexpensive and easy to make. Good parts can be cast from them.

REVIEW

1. Describe two advantages and uses for latex molds.

2. Describe a latex elastomer mold.

3. Name two latex mold pattern materials.

4. Describe each step in latex mold making.

5. Explain the meaning of RTV.

6. List each step in preparing a pattern for latex coating.

7. Describe the procedure and explain the reason for making a latex mold back-up.

8. Name two materials that can be cast into latex molds to make a product.

SELECTED BIBLIOGRAPHY

Flexible Mold Making—Teacher's

Manual. Indianapolis: Howard W. Sams & Company, Inc., 1974.

Richardson, Terry A. *Modern Industrial Plastics.* Indianapolis: Howard W. Sams & Company, Inc., 1974.

Roff, W.J. and Scott, J.R. *Handbook of Common Polymers, Fibers, Films, Plastics and Rubbers.* Cleveland: CRC Press, 1971.

EQUIPMENT AND MATERIAL SUPPLIERS

1. Ain Plastics, Inc., 65 Fourth Avenue, New York, New York 10003.

2. Delvies Plastics, Inc., 2320 South West Temple, P.O. Box 1415, Salt Lake City, Utah 84110.

3. Industrial Arts Supply Company, 5724 West 36th Street, Minneapolis, Minnesota 55416.

Latex mold making equipment and materials can be purchased from any of the suppliers listed above.

Assignment 13
Epoxy Molds

OBJECTIVES

To prepare a flat back pattern.

To make an epoxy mold.

INTRODUCTION

An epoxy mold can be a cavity or a positive type mold. Plastics are either placed in or over the mold to form into a product. Epoxy mold material is a composite. A composite is made of about 5% to 20% epoxy resin and about 80% to 95% metal by weight. Some metals used are steel, aluminum, bronze, zinc, and stainless steel. Metals add bulk, strength, density, and heat dissipation properties. The epoxy acts as the binder material for the composite. These composites are sold as liquids and putties.

Epoxy mold materials are used in the plastics industry. They are used for vacuum forming molds, embossing dies, low production injection molds, urethane foam casting molds, slush vinyl dispersion molds, and reinforced plastics molds. Epoxy molding materials have a fairly long life.

The epoxy mold making material is a two-part system. An easy epoxy system to work with is the room temperature curing one. This type was used in this assignment.

Often an epoxy mold will be exposed to high temperatures. A special heat-resistant epoxy system is used. This material requires special resin heating steps. Also, it requires a special oven curing cycle.

Epoxy molding materials are fairly economical. They are also easy to work with in a school laboratory.

SAFETY

The following precautions should be taken when making epoxy molds:

1. Work in a well-ventilated area. Do not breathe the resin and chemical hardener fumes. People with asthma must *never* breathe these fumes.

2. Wear safety glasses and disposable gloves.

3. Protect your eyes and skin from contact with epoxy resins and hardener. If contact is made, *immediately* wash your skin and flush your eyes with clean water for 15 minutes.

4. Wear a respirator when machining epoxy molds.

5. Turn on the dust removal system before machining epoxy molds.

EQUIPMENT AND MATERIALS

The equipment and materials needed for making epoxy molds are:

1. Safety glasses.
2. Disposable (polyethylene) gloves.
3. Respirator (for machining epoxy molds).
4. Retainer box.
5. Pattern.
6. Plywood molding board.
7. Epoxy resin and hardener.
8. Release agent (supplied by epoxy manufacturer).
9. Rubber cement.
10. Modeling clay.
11. Fillet iron.
12. Bulking materials (alumium chips, steel shavings, or wood flour).
13. Unwaxed cups for mixing.
14. Mixing sticks.
15. Vacuum degassing equipment.
16. Measuring rule.
17. Weight scale.

Epoxy Molds

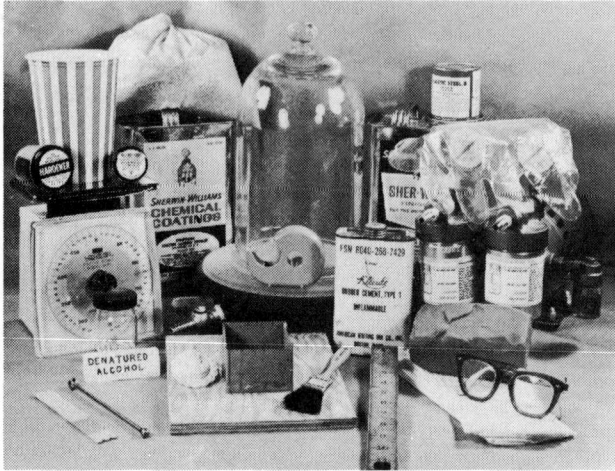

Fig. 13-1 Epoxy mold making equipment and materials.

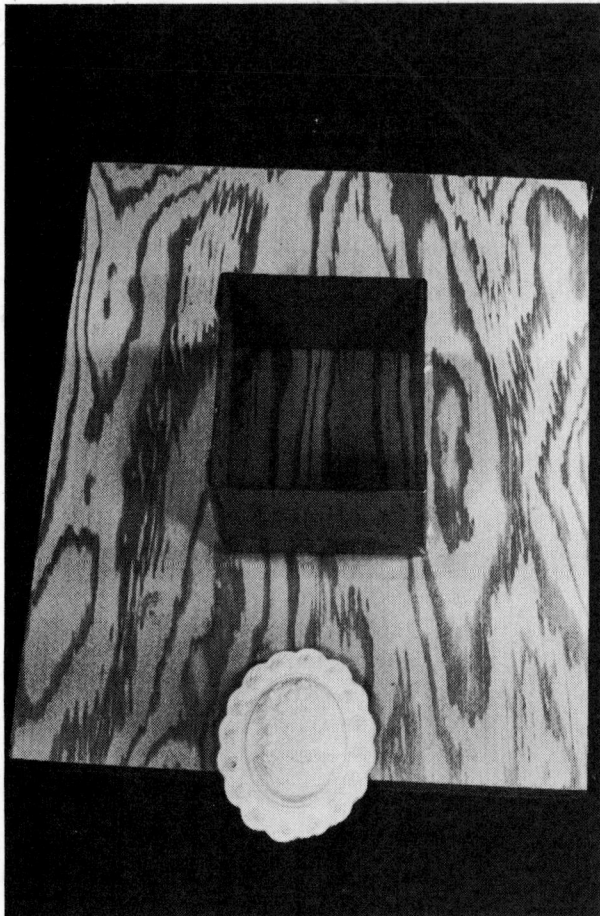

Fig. 13-2 Pattern and retainer box.

18. 1 one-inch brush.
19. Denatured alcohol.
20. Lacquer.
21. Lacquer thinner.
22. Clean cloth or paper towels.
23. Cellophane tape.

EPOXY MOLD LAB PROCEDURE

1. Obtain all the items listed under "Equipment and Materials" needed to make an epoxy mold. See Fig. 13-1.

2. Obtain the pattern and retainer box. See Fig. 13-2. Patterns are made of high density plaster, wood, or metal. The pattern should have no undercuts. Retainer boxes are made of sheet plastic, sheet glass, or sheet metal. Make sure the retainer box is taped together and taped to the molding board.

3. Place rubber cement on the pattern back. See Fig. 13-3. This stops the pattern from floating when epoxy is cast around it.

4. Center the pattern in the retainer box. The distance of the pattern from the retainer box determines the mold wall thickness. This distance should be ½". Make sure the pattern lies flat on the molding board.

5. Place a fillet of modeling clay around the pattern bottom and on all inside retainer box corners. See Fig. 13-4. Make the fillet with a fillet iron. The fillets stop the cast epoxy from running under the pattern and forcing air bubbles out. They also cause a smooth outside corner to be cast in epoxy.

6. Brush two coats of lacquer on porous unfinished wood or plaster patterns. See Fig. 13-5. This stops the pattern from releasing trapped gas into the epoxy.

7. Apply several coats of release

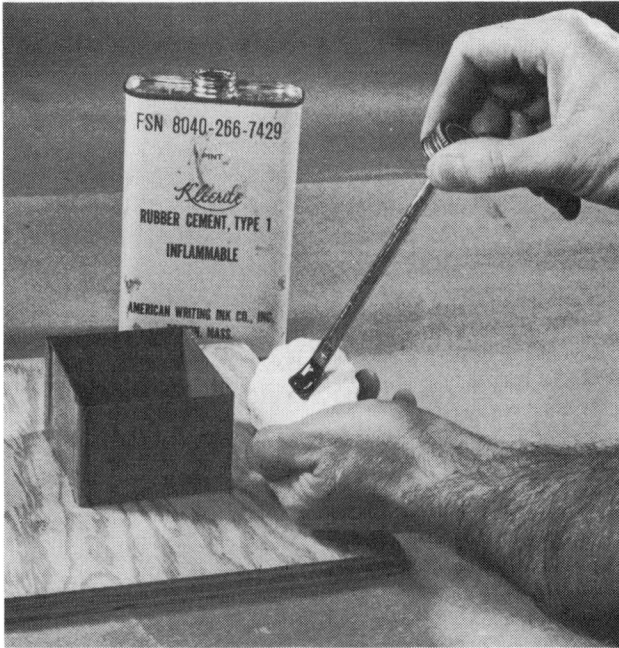

Fig. 13-3   Applying rubber cement.

Fig. 13-4   Placing fillet of modeling clay.

agent to the pattern, molding board, and retainer box. Let each coat dry and then gently buff it before applying the next one.

8. Determine the volume of the retainer box. See Fig. 13-6. This is done by first measuring the width, depth, and length of the box. Then, multiply them together. Now, estimate the volume of the pattern and subtract it from the retainer volume figure. This answer is the volume to be filled.

9. There is another way to estimate the volume of the retainer box. Fill the retainer box with water. Next, pour the water into a volume calibrated (marked) container. This tells you the size of the space (volume) taken up by the retainer box. Be sure to dry the retainer box before going on to the next step.

10. Determine the amount of epoxy needed to fill the calculated retainer box volume. This is done by checking the epoxy manufacturer's weigh-to-volume ratio specification. It will differ for each epoxy type. The specification shown states that one pound of epoxy will fill 11.4 cubic inches of volume. See Fig. 13-7. If the volume of the retainer box is 22.8 cubic inches, two pounds of epoxy must be mixed (22.8 divided by 11.4 = 2). There will be extra epoxy because the pattern takes up some volume in the retainer box.

11. Weigh the amount of resin and hardener needed. See Fig. 13-8. Your instructor or epoxy specification sheet will tell you the amount of hardener to use. The specification shown asks you to mix one part hardener to 9 parts resin by weight. The resin-to-hardener ratio will differ for each type of epoxy.

12. Pour the hardener into the container of resin. See Fig. 13-9. Use

Epoxy Molds

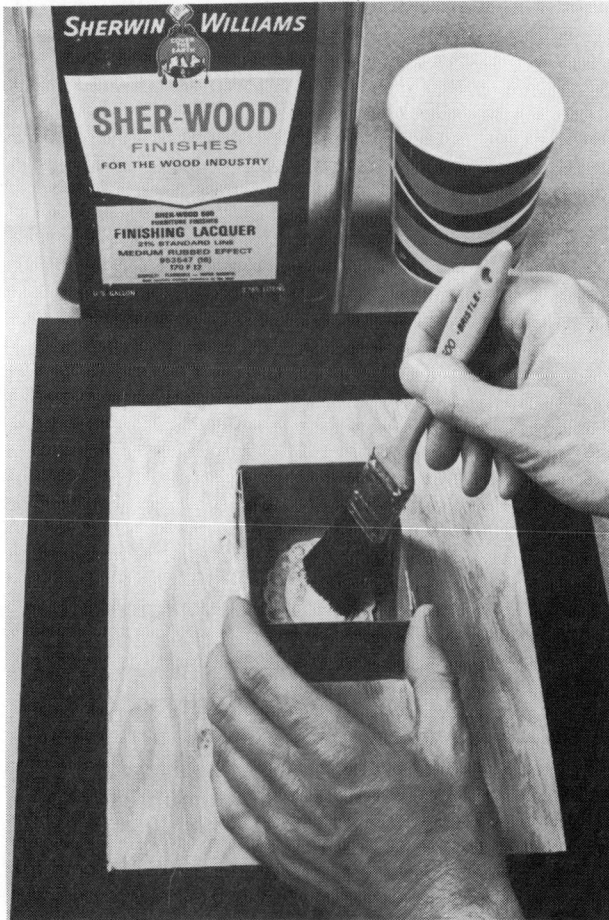

Fig. 13-5　Applying lacquer to pattern.

Fig. 13-6　Calculating volume of retainer box.

an oversized unwaxed paper container. A large container will let you vacuum degas the epoxy. Use disposable gloves when handling epoxy parts.

13. Mix the two components well. Try not to trap air in the material. Be sure to scrape the side and bottom of the container with the stirrer. This helps completely mix the epoxy parts. Your instructor will tell you how long to mix the epoxy parts before they start to harden (pot life).

14. Degas the epoxy with a vacuum degassing unit. See Fig. 13-10. Make sure the container is about four times larger than the mixed epoxy. The escaping air will make the epoxy rise. Degassing is finished in about two minutes or when the foaming stops. Use about 28 to 29 inches of mercury of vacuum. **Do the next step only if the epoxy was not degassed.**

15. Brush a bubble-free coat of epoxy on the pattern. See Fig. 13-11. This helps remove air bubbles from the pattern surface. Be sure to clean the brush in denatured alcohol immediately.

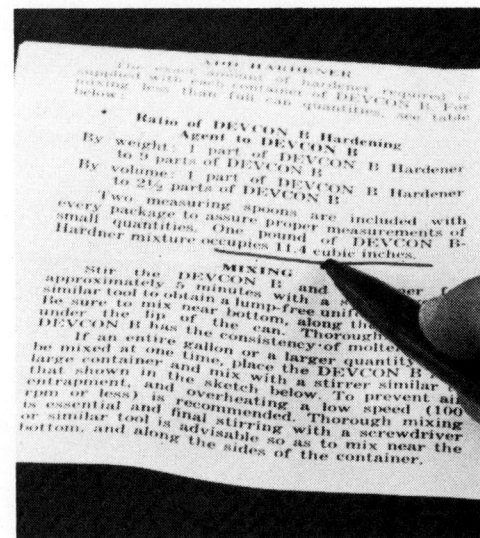

Fig. 13-7　Checking weight-to-volume ratio.

Section III: Moldmaking

Fig. 13-8 Weighing out resin.

Fig. 13-9 Adding hardener.

Fig. 13-10 Degassing the epoxy.

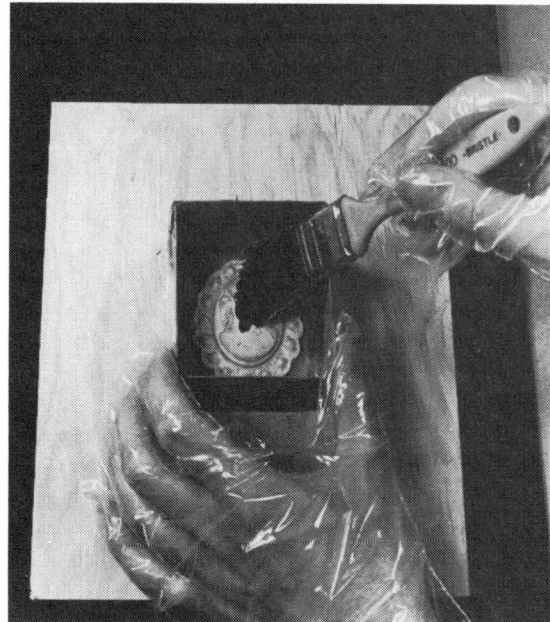

Fig. 13-11 Applying epoxy coat to pattern.

Fig. 13-12   Pouring epoxy into retainer box.

Fig. 13-13   Mixing bulking agent with catalyzed epoxy.

16. Pour the epoxy into a corner of the retainer box. See Fig. 13-12. Let it slowly flow over the pattern. The epoxy should be poured until it is about ½″ over the highest part of the pattern. The pouring method shown helps stop air from being trapped against the pattern surface.

17. Let the epoxy cure overnight. The instructor will tell you the cure time. If high temperature epoxy is used, it may require a special oven cure. Note that Steps 18, 19, and 20 are optional for use with big molds.

18. Mix a bulking agent into catalyzed epoxy. See Fig. 13-13. The bulking agent can be aluminum chips, wood flour, or steel chips. Bulking agents increase the volume of the epoxy. They are used to fill large retainer box areas. They also reduce the cost of the epoxy mold.

19. Pour the epoxy bulking agent over the almost-cured first epoxy layer. See Fig. 13-14. Let the first layer almost cure before pouring a bulking epoxy. This stops the bulking agent from sinking through the first layer and into the pattern.

20. Let the epoxy cure overnight. High temperature epoxy may require an oven cure. Be sure to clean all epoxy covered tools in denatured alcohol. Clean your hands with denatured alcohol, too.

21. Remove the retainer box tape.

22. Lift the mold off of the pattern and turn it over. See Fig. 13-15.

23. Check the epoxy mold. See Fig. 13-16. If it has soft spots, the epoxy parts were not well mixed. Air bubble holes in the mold surface can be filled with more catalyzed epoxy. Use a toothpick or an artist brush. The mold can now be set up to be used as a thermoforming mold.

Fig. 13-14   Pouring epoxy bulking agent.

Fig. 13-15   Lifting mold off pattern.

Fig. 13-16   Checking epoxy mold.

Several epoxy molds are shown in Fig. 13-17. One-piece epoxy molds are often used as thermoforming molds. Two-piece epoxy molds are often used as injection and blow molding molds. The two-piece molds are made as shown in the "Silicone Moldmaking" assignment.

CONCLUSION

Most epoxy materials cure through chemical reactions between the resin and the hardener. During the reaction, heat (exothermic) is produced. A large epoxy mold curing makes a lot of heat. It also cures faster than one that has a small mass. A casting larger than four inches thick should be cast in layers of 1 ½". To stop casting heat build-up, warpage, and shrinkage, use an 8 to 1 span-to-depth casting ratio. Also, each layer should cure ½ hour before pouring the next layer. The average cure time is about three hours.

To stick epoxy mold material to itself or to other materials, do the following. Remove oil, dirt, rust, paint, moisture, and release agents from the surfaces. Let the material reach room temperature and roughen (sand) it. Then, apply the epoxy.

Each epoxy molding system has its advantages. Generally epoxy molds are used as prototype molds. Choosing the proper epoxy system requires knowledge of different epoxy properties.

REVIEW

1. Give two uses for epoxy molds.

2. Describe an epoxy mold.

3. Name two epoxy mold pattern materials.

125

Epoxy Molds

4. Describe each step in epoxy mold making.

5. Describe each step in preparing a pattern and a retainer box.

6. Name two plastics that can be molded with epoxy molds.

7. Name two metals used as part of the epoxy molding material composite.

8. Name the type of epoxy molding system you used.

9. Describe how to make two epoxy surfaces stick together.

10. Why should an 8-to-1 span-to-depth casting ratio be used?

11. Explain why a fillet is placed around the pattern and the retainer box corners.

12. Explain how the epoxy volume-to-weight ratio and the resin-to-hardener ratio are calculated.

13. List three ways to stop epoxy mold air entrapment.

14. Explain the reason for using an epoxy bulking agent.

15. Name two advantages of epoxy molds.

16. Explain what happens to the finished epoxy mold when the resin and hardener are not well mixed.

17. Describe two epoxy mold curing methods.

Selected Bibliography

Baird, Ronald J. *Industrial Plastics.* South Holland, Illinois: The Goodheart-Willcox Company, Inc., 1971.

Fig. 13-17 Examples of several epoxy molds and products.

Cast Epoxy Tooling—Teacher's Manual. Indianapolis: Howard W. Sams & Company, Inc., 1974.

Devcon Corporation. *Devcon B.* Bulletin D-2. Danvers, Massachusetts: Devcon Corporation, 1974.

——. *Devcon C.* Bulletin D-3. Danvers, Massachusetts: Devcon Corporation, 1968.

——. *Devcon C-2.* Bulletin D-16. Danvers, Massachusetts: Devcon Corporation, 1968.

Milby, Robert V. *Plastics Technology.* New York: McGraw-Hill Book Company, 1973.

Patton, William J. *Plastics Technology: Theory, Design, and Manufacture.* Reston, Virginia: Reston Publishing Company, Inc., 1976.

Richardson, Terry A. *Modern Industrial Plastics.* Indianapolis: Howard W. Sams & Company, Inc., 1974.

Rosato, Dominick V., ed. *Plastics Industry Safety Handbook.* Boston: Cahners Books, 1973.

EQUIPMENT AND MATERIAL SUPPLIERS

1. Brodhead-Garrett, 4560 East 71st Street, Cleveland, Ohio 44105.

2. Cope Plastics, Inc., 4441 Industrial Drive, Godfrey, Illinois 62035.

3. Delvies Plastics, Inc., 2320 South West Temple, P.O. Box 1415, Salt Lake City, Utah 84110.

4. Industrial Arts Supply Company, 5724 West 36th Street, Minneapolis, Minnesota 55416.

Epoxy Molds

5. McKilligan Industrial Supply Corporation, 494 Chenango Street, Binghamton, New York 13901.

6. Paxton/Patterson, 5719 West 65th Street, Chicago, Illinois 60638.

Epoxy mold making equipment and materials can be purchased from any of the suppliers listed above. For degassing units, see the suppliers listed in Assignment 2 (Thermosets).

Section IV
Fabricating
Plastics

Assignment 14
Chemical Welding

OBJECTIVES

To prepare plastic for chemical welding.

To chemically weld lap, tee, and butt joints.

INTRODUCTION

Chemical welding is an industrial solvent cementing process. Solvent is placed on the areas of the plastic being joined. This solvent softens these areas. The softened areas are pressed together and they join. After the solvent evaporates (drys), the chemical welding is finished.

Solvent cement will often contain dissolved pieces of plastic. This makes the solvent thick. This cement is used to chemically weld plastics. Also, it is used to fill low spots and nicks in the plastic being cemented. This type of solvent cement is called "bodied" or "dope" cement.

Chemical welding is generally done by the capillary and the dip process. These processes are demonstrated in this assignment.

SAFETY

The following precautions should be taken when chemical welding:

1. Work in a well-ventilated area. Do not breathe solvent cement fumes.

2. Wear safety glasses.

3. Keep the solvent cement away from flames.

4. Do not swallow solvent cement.

5. Protect your eyes and skin from contact with solvent cement. If you splash cement in your eyes, wash them *immediately* with clean water. Then, *get medical attention.*

EQUIPMENT AND MATERIALS

The equipment and materials needed for chemical welding are:

1. Safety glasses.
2. Acrylic, cellulose acetate, cellulose nitrate, nylon, polycarbonate, or general purpose polystyrene.
3. Methylene chloride (MDC), ethylene dichloride (EDC), or 1-1-2 trichlorethane solvent cement.
4. Cement applicator.
5. Steel scraper.
6. Course, medium, and fine wet/dry sandpaper.
7. Try square.
8. Measuring scale.
9. Woodworking bench vise.
10. Cementing fixture.
11. Abrading block.
12. Tray or jar.
13. Clean cloth or paper towels.

CHEMICAL WELDING LAB PROCEDURE

1. Obtain all the items listed under "Equipment and Materials" needed to chemical weld. See Fig. 14-1.

2. Obtain the proper solvent cement for the plastic to be welded. Acrylic is chemically welded in this assignment. Methylene chloride (MDC), Ethylene Dichloride (EDC) or 1-1-2 Trichlorethane can be used to chemically weld acrylic. Most plastics that can be chemically welded require a specific solvent cement. Ask your instructor what cement to use for different plastics. Table 14-1 lists a few of the plastics that can be chemically welded (solvent cemented).

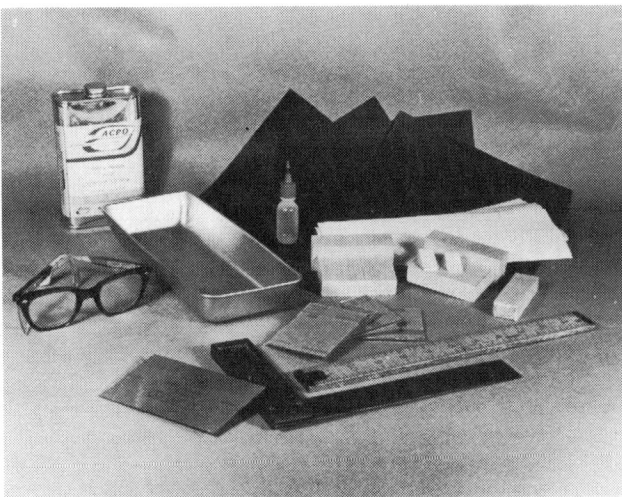

Fig. 14-1 Chemical welding equipment and materials.

Chemical Welding

3. Cut the plastic to size. Ask your instructor for plastic sizes.

4. Put each piece of plastic in a woodworking vise. Be sure the edge to be cemented is up. Leave the masking paper on the plastic.

5. Scrape all plastic edges that will be cemented with a woodworking scraper. See Fig. 14-2. Scrape until all of the cutting marks are removed.

	Acetal	Acrylics	Cellulose acetate	Cellulose acetate butyrate	Cellulose nitrate	Ceramics	Glass	Metal	Nylon	Phenolic	Poly-carbonate	Styrene	Urea	Vinyl, flexible	Vinyl, rigid	Wood
Acetal	E,R	B,E,R	R,E	R,E	R,E,	E,R	E,R	E,R	E,R	E,R	E,R	E,R	B,E,R	S,B,R	B	E,R
Acrylics	B,E	M,S,B	B,E	B,E	B,E	B,E,R	R,E,B	E,R,B	R,E,B	E,R,B	B,R,S	S,R,B	E,R			B,R,E
Cellulose acetate	R,E	E,B	S,B	B	B,S	B	R	R,E	R,E	R,E	B,R	R	E,R	R	E	B
Cellulose acetate butyrate	R,E	B	B	B	B	E,R	E,R	E,R	R,E	E,R	B,R	E,R	E,R	R	E,R	B
Cellulose nitrate	R,E	R,E,S	S,B	S,B	S	B,R	B,R	R,E	R,E	R,E	R,E	E,B,R	E,R	R	R,E	B
Nylon	E,R	R,E	R	R	R	E,R,S	R,E	R,E	S,B	S	R,E	R	R,E	R	R	R,E
Phenolic	E,R	E,R	R,E	R,E	R,E	E,R	E,R	E,R	S,E	S,E	E	R,E	R	R,E	R,E	E,R
Poly-carbonate	E,R	M,R,E	E,R	E,R	E,R	E,R	E,R	E,R	E,R	E,R	S,E,R	E,B	E,R	R,E	E,R	E,R,B
Polyester	E,R	E	E	E	E	E	E	E	E	E	E	E	E	E	E	E
Styrene	E,R	R,B	R	R	R	R,B	R,B	B,R	R,B	R,B	R,B	R	R,B	R,B	R,B	R,B
Vinyl, flexible	B	S,B,R	B	B,R	B,R	B,R	B,R	R,B,E	R,E	R,E	B	R,E	E,R	S,R,B	S	B,R
Vinyl, rigid	B	B,R	R,B	B	B,R	B,R	R,B	R,E,B	B,R	E,R	R,B	B,R	E,R	B,R,S	S,B	R,B

Abbreviations and definitions:
M—Monomeric cements, based on a specific plastic which must be catalyzed to produce a strong bond.
S—Solvent cements, which dissolve the plastic to provide molecular interlocking, then evaporate. Normally they require close fitting joints, produce strong bonds.
B—Bodied adhesives contain a thermoplastic or thermosetting resin, and solvents, sometimes plasticizers, which dry by evaporation. The bodied adhesives can compensate for substantial variations in mating surfaces while still providing strong joints.
R—Elastomeric adhesives, based on natural or synthetic rubbers. Some contain the rubber in solvent or water suspension or solution. They may be cured at room or elevated temperatures to provide extremely strong joints.
E—100% reactive adhesives which depend upon catalytic action to join the two materials with an interlayer of thermosetting resin. The various dry and liquid epoxies, the polyesters, and the phenolics are being considered in this category.

Polyethylene can be bonded only by some of the rubber based cements with extremely low joint strengths.
Fluorocarbons can be bonded to themselves and to other materials only if pretreated or etched. Normally, epoxy adhesives are used. Joint strength is moderate.

Table 14-1 Plastics and recommended adhesives. (Reprinted by permission of *Modern Plastics Magazine*, McGraw-Hill, Inc.)

Section IV: Fabricating Plastics

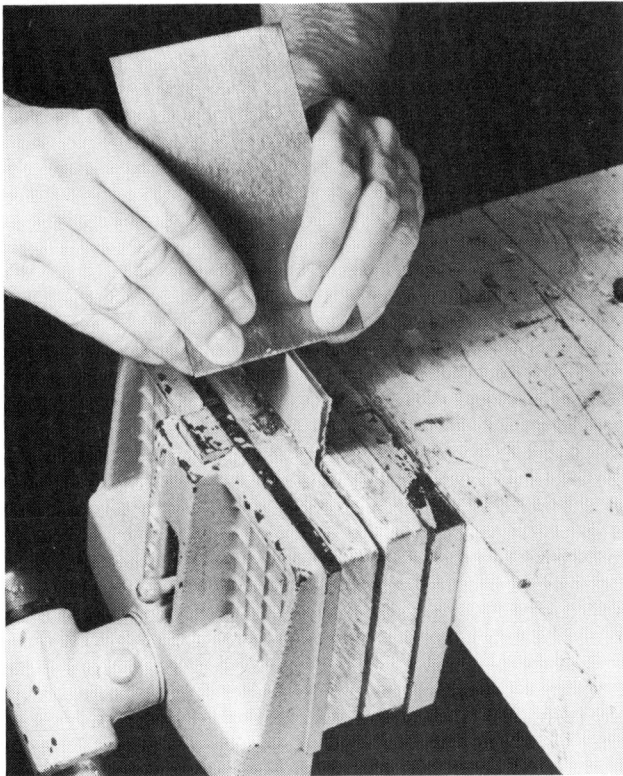

Fig. 14-2 Scraping plastic edges.

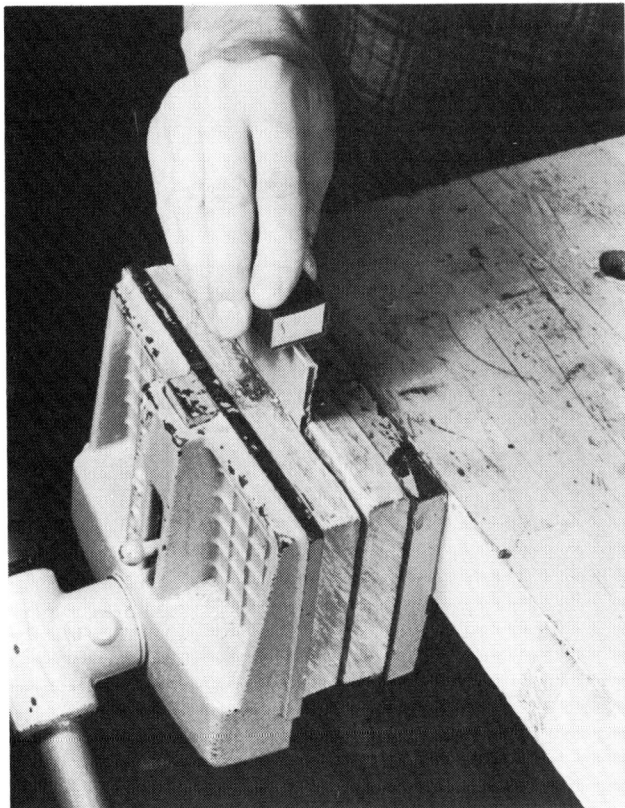

Fig. 14-3 Sanding the edges.

6. Sand each edge to be cemented. See Fig. 14-3. Sand the plastic edge until it has a frosted finish. Begin with 100 grit, then 220 grit, and finish with 400 grit. Be sure to use a sanding block for all sanding. This helps prevent the rounding off of the plastic edges.

7. Peel the masking paper off of all the plastic to be cemented.

8. Wipe all dust from all edges to be cemented.

9. Dip each edge to be cemented into solvent cement. See Fig. 14-4. Leave the edges in the cement until they soften and slightly swell. This is called *dip chemical welding.*

10. Lay the plastic in the fixture with the wet edges facing each other.

11. Press the wet edges firmly together until the cement drys.

12. Clamp the edges tightly with the fixture wedges for 5 to 10 minutes. See Fig. 14-5. The result is a welded butt joint.

13. Place one plastic edge over another plastic edge to form a lap joint. The top plastic edge should overlap the bottom piece by ¼".

14. Squeeze the cement applicator while moving it along the joint. See Fig. 14-6. The cement will be drawn between the plastic and into the joint. This process is called *capillary cementing.* Be sure to keep finger pressure on the joint until the cement dries.

15. Place light pressure on the dry joint for 5 to 10 minutes. Use a clamp or a weight to place pressure on the joint.

16. Hold one abraded plastic edge at right angles to a flat piece of plastic. This is a "T" joint.

Chemical Welding

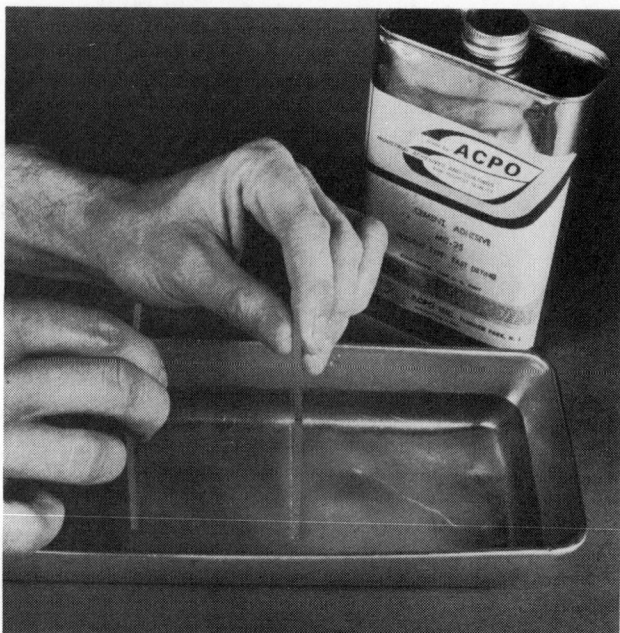

Fig. 14-4 Dipping edges into cement.

Fig. 14-5 Clamping edges.

17. Repeat Steps 14 and 15 for the "T" joint. See Fig. 14-7. Make sure the joint is square by checking it with a try-square.

18. Check the joint weld line for air bubbles. If found, force the cemented plastic pieces together. The extra pressure will drive the air out.

19. Check the weld strength. If the joint is easily broken, use more cement and pressure. Also, keep the joint under pressure longer. If the plastic is dry and does not stick, work faster and use more cement.

20. Check the plastic along the weld line. If it is melted in the weld area, use less cement and pressure.

21. Note the weld line of the joint shown in Fig. 14-8. This is a good joint. It is clear, smooth, and free of air bubbles.

CONCLUSION

Many products joined by chemical welding are shown in Fig. 14-9. These products are made of acrylic, general purpose styrene, and polycarbonate.

Chemical welding is used by both industry and home craftsmen. Many household products are assembled with solvent cements. These products include record player dust covers, frame tables, display book stands, serving trays, model covers, and game boards.

Many plastics can be chemically welded or joined together with solvent cement. Listed in Table 14-1 are many plastics that can be solvent cemented. Also listed are many other types of cementing techniques.

Fig. 14-6 Moving cement applicator along joint.

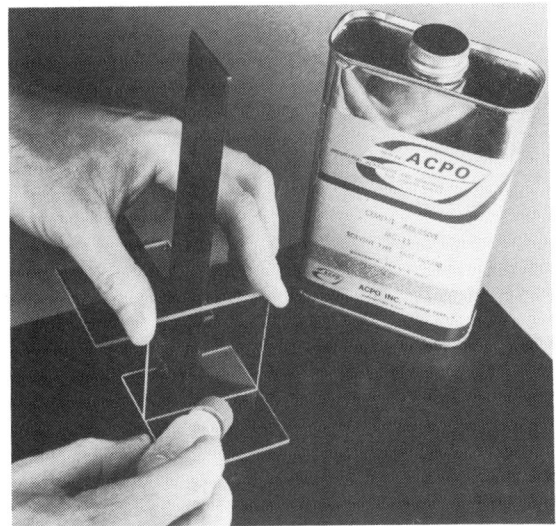

Fig. 14-7 Forming "T" joint.

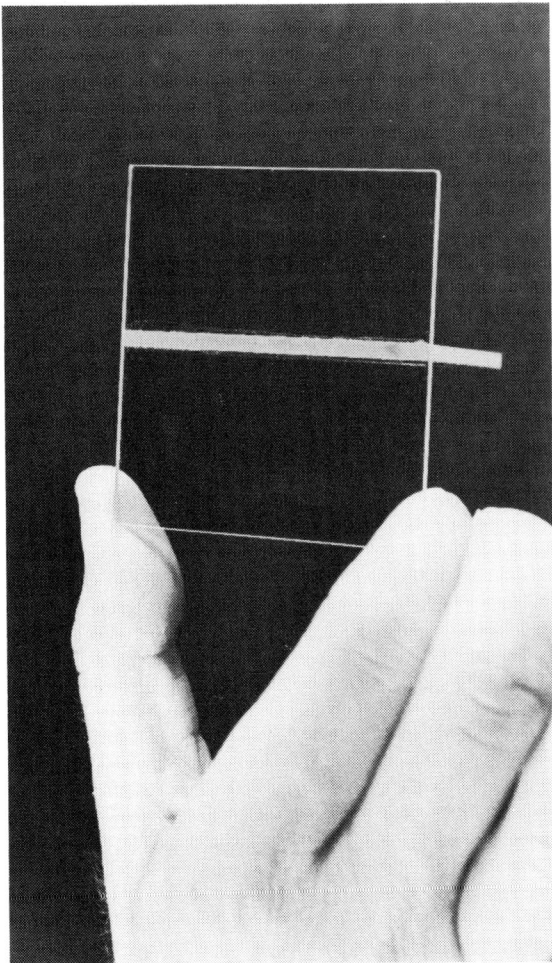

Fig. 14-8 Weld line of good joint.

Fig. 14-9 Examples of products joined by chemical welding.

Chemical Welding

REVIEW

1. Name the plastic you chemically welded.

2. Explain how chemical welding works.

3. List four products that have been put together by solvent cementing.

4. Give the chemical name for the solvent cement you used.

5. Name a use for bodied cement.

6. Describe how capillary cementing works.

7. Sketch a lap joint and show where a chemical weld should be made.

8. Explain the reason for the masking on plastic sheet.

9. List three characteristics of a good chemical welded joint.

SELECTED BIBLIOGRAPHY

Agranoff, Joan, ed. *Modern Plastics Encyclopedia.* New York: McGraw-Hill Book Company, 1976-77.

Baird, Ronald J. *Industrial Plastics.* South Holland, Illinois: The Goodheart-Willcox Company, Inc., 1971.

Milby, Robert V. *Plastics Technology.* New York: McGraw-Hill Book Company, 1973.

Patton, William J. *Plastics Technology: Theory, Design, and Manufacture.* Reston, Virginia: Reston Publishing Company, Inc., 1976.

Richardson, Terry A. *Modern Industrial Plastics.* Indianapolis:

Howard W. Sams & Company, Inc., 1974.

Warner Electric Company. *How to Do Plastic Laminating for Pleasure and Profit.* Chicago: Warner Electric Company, Inc., 1976.

EQUIPMENT AND MATERIAL SUPPLIERS

1. Ain Plastics, Inc., 654th Avenue, New York, New York 10003.

2. Brodhead-Garrett, 4560 East 71st Street, Cleveland, Ohio 44105.

3. Cadillac Plastics & Chemical Company, P.O. Box 810, Detroit, Michigan 48232.

4. Cope Plastics, Inc., 4441 Industrial Drive, Godfrey, Illinois 62035.

5. Delvies Plastics, Inc., 2320 South West Temple, P.O. Box 1415, Salt Lake City, Utah 84110.

6. Graves-Humphreys, Inc., 1948 Franklin Road, P.O. Box 1347, Roanoke, Virginia 24033.

7. Industrial Arts Supply Company, 5724 West 36th Street, Minneapolis, Minnesota 55416.

8. Paxton/Patterson, 5719 West 65th Street, Chicago, Illinois 60638.

The plastics and solvent cements used in chemical welding can be purchased from any of the above listed suppliers. Applicators can be purchased from suppliers 1, 3, 4, 5, and 7.

Chemical Welding

Assignment 15
Hot Air Welding

OBJECTIVES

To select the proper hot air welding materials.

To prepare the materials for hot air welding.

To make a hot air welded joint.

INTRODUCTION

Hot air welding is a method of joining thermoplastics. Hot air welded products include industrial tanks, containers, funnels, ducts, pipes, and hoods.

Hot air welding works as follows. Gas or air is heated as it flows through a hot welding gun. This hot gas is made to flow over thermoplastic materials and welding rod. The hot gas softens the plastics and rod. Then, the softened plastics are either forced together or joined with the soft rod. The welded joint is then cooled.

Certain thermoplastics may crack after they have been welded. This is called *post-weld stress cracking.* Often these plastics can be welded in a nitrogen gas atmosphere. This helps stop weld stress cracking. Nitrogen, rather than air, is fed to the welding gun and heated. Polyethylene and polypropylene are two plastics often welded in a nitrogen atmosphere.

Skill and practice are needed by the welder to make good welds. The welder must know the correct welding temperature, pressure, rod angle, and speed for each weld. The welding temperatures, pressure, rod angle, and speed are different for each type and size of plastic. They are also different for each rod type, shape, and size.

SAFETY

The following precautions should be taken when hot air welding:

1. Work in a well-ventilated area. Do not breathe plastic fumes because some are toxic.

2. Wear safety glasses and heat-resistant gloves.

3. Work on a heat-resistant surface.

4. Keep a general purpose fire extinguisher in the work area.

5. Obey all plastic suppliers' safety rules for handling the plastics. Some plastics make noxious and poisonous fumes when welded.

6. Do not burn yourself with the hot air from the welder.

7. Do not touch hot plastic, welding tip, or barrel.

8. Keep your hands and fingers out of the welding zone.

9. Keep the hot welding torch in its holder when not welding.

10. Make sure the welding gas is flowing through the torch **before** you connect to the torch.

11. Let the welding gas flow through the torch for five minutes **after** you unplug electricity.

EQUIPMENT AND MATERIALS

The equipment and materials needed for hot air welding are:

1. Safety glasses.
2. Heat-resistant gloves.
3. Heat-resistant surface.
4. Welding unit.
5. Compressed air or nitrogen lines.

Hot Air Welding

Fig. 15-1 Hot air welding equipment and materials.

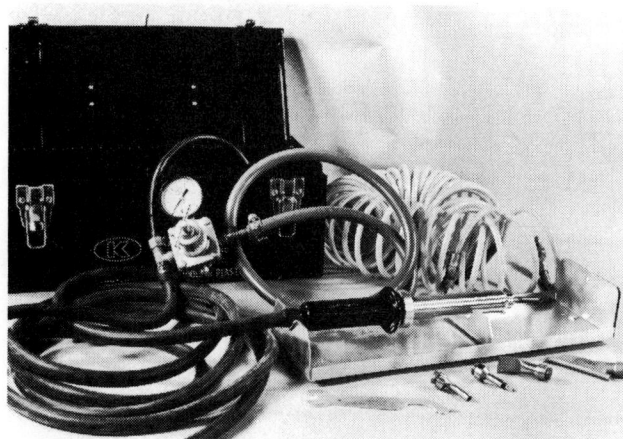

Fig. 15-2 Hot air welding unit.

Fig. 15-3 Cutting plastic for a lap joint.

6. Low pressure hot gas torch regulator.
7. Compressed air or nitrogen regulator.
8. Round, oval, flat, or triangular welding rod of the proper size and material.
9. Welding tips (general-purpose, tacking, flat, triangular, high speed for corner, round, or flat strips).
10. PVC, polyethylene, or polypropylene plastic.
11. Holding fixtures for plastic (optional).
12. Abrasive paper.
13. Powered abrasive wheel.
14. Knife or wire cutters.
15. Weights (optional).
16. Torch stand.
17. 2 C-clamps.
18. Cloth or paper towels.

HOT AIR WELDING LAB PROCEDURE

1. Obtain all the items listed under "Equipment and Materials." See Fig. 15-1.

2. Obtain the welding unit needed to hot air weld a product or joint. See Fig. 15-2.

3. Select the proper welding rod. Rods are sold in many shapes. The most common shapes are round, flat, and triangular. Each rod shape is sold in different colors, sizes, and thermoplastics. Make sure a welding torch tip is used that matches the rod. Homemade welding rods are made by cutting a thin strip from the plastic to be welded. Ask your instructor what sizes to make the homemade rods.

4. Cut two equal size pieces of plastic to make a lap joint. See Fig. 15-3. Ask your instructor what sizes to cut the plastic. Many thermoplastics can be hot gas welded. Some are acrylic,

Section IV: Fabricating Plastics

Fig. 15-4 Clamping the overlapped sheets.

Fig. 15-5 Inserting a heater element.

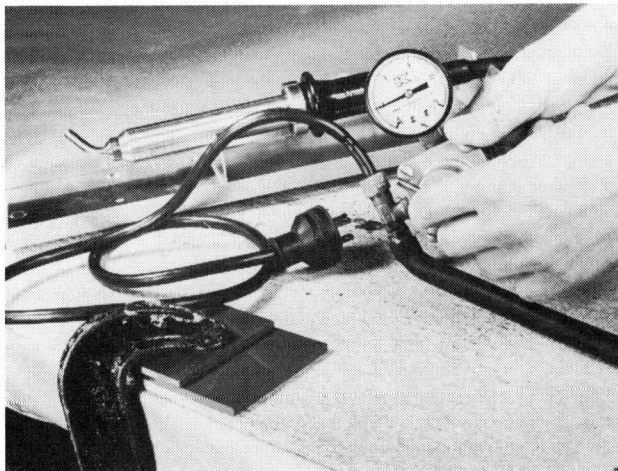

Fig. 15-6 Adjusting welding air regulator.

polycarbonate, polysulfone, acetal, polyvinyl chloride, ABS, nylon, polyethylene, polyphenylene, and chlorinated polyethers. The material used in this assignment is PVC.

5. Wipe the surfaces to be welded. Make sure the surfaces are free of oil, dirt, grease, and mold release. **Do not use a solvent.** It may soften the plastic and weaken the weld joint.

6. Overlap the two sheets of plastic about 1¼." See Fig. 15-4. Be sure to clamp them in place.

7. Place the proper heating element in the welding unit or set the heat adjustment switch. See Fig. 15-5. Your instructor will tell you what element or switch adjustment to use.

8. Adjust the welding air regulator to 2½ to 3½ psi. See Fig. 15-6. This lets compressed air flow through the torch. Ask your instructor for the proper adjustment.

9. Plug in the electrical heating element. The heat from the element heats the compressed air flowing around it. **Be sure air is flowing before you plug in the heating element.** Lower the air pressure, place a higher rated element in the welder, or adjust the heat switch higher if more heat is needed.

10. Pass hot gas over the plastic where the weld is to start. Keep the welding tip about ½" from the plastic surface. See Fig. 15-7. Use a fanning (2 oscillations per second) motion to spread heat over the plastic. Be sure about ½" of the rod and ⅜" of the plastic are being heated. The heat is applied in the direction of the weld.

11. Move the welding rod (60° angle cut on end) up and down while touching the hot plastic. See Fig. 15-8. Heat the plastic and rod with a fanning motion. About 60% of the

Hot Air Welding

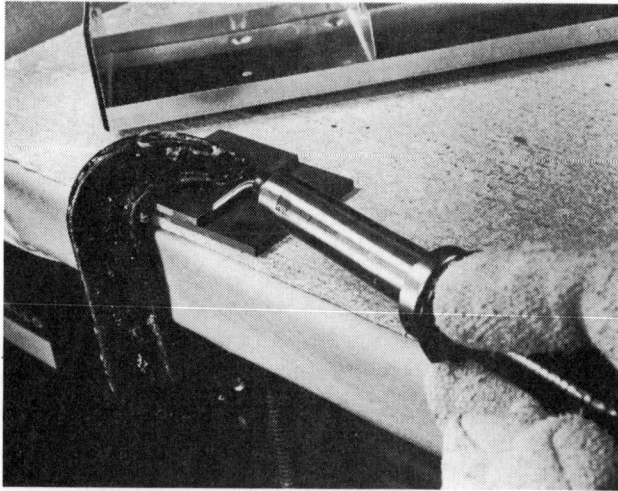

Fig. 15-7 Applying hot gas to weld area.

Fig. 15-8 Welding rod movement.

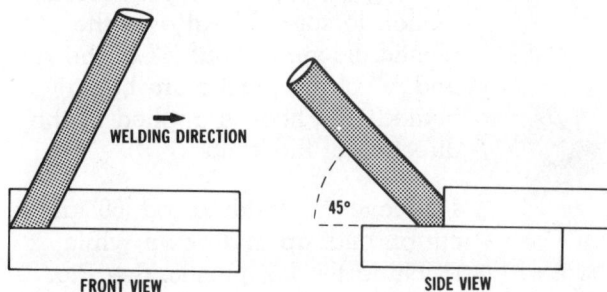

WELDING DIRECTION

45°

FRONT VIEW   SIDE VIEW

Fig. 15-9 Welding rod angle.

heat should be directed to the plastic and 40% to the rod.

12. Stick the rod firmly into the shiny plastic using light pressure. Angle the rod a little in the welding direction. See Fig. 15-9. Keep supplying heat to the rod and plastic as described in Step 11. Hold the rod so that it is at a 45° angle to the two welded surfaces. Also, hold the rod 90° to the root (bottom) of the weld. See Fig. 15-10.

13. Weld about ½" as described in Step 12 and then slightly slant the rod back. See Fig. 15-11. Do **not** slant the rod to the left or right of the bead. See Fig. 15-12. Heat the rod and plastic as described in Step 11. Weld about 6 to 8" per minute. Make sure you use only enough steady pressure to fuse the rod to each plastic. Too much pressure will stretch the rod. It may crack later. **Do not let the rod lift and trap air under it.** Trapped air causes a weak weld.

14. Remove the torch. Twist, cool, and cut off the rod. See Fig. 15-13. Use a knife or wire cutters to cut the rod.

15. Check the weld. There should be even flow lines along the rod edges. Also, a wave of semi-molten plastic should be in the base plastic. This wave will be in front of the rod while welding. If the weld is cold, no flow lines will be seen. The joint will be weak. If the weld is too hot, brown or black areas will show on PVC. Also, uneven flow lines will show on the rod edges.

16. Unplug the electric from the welder element. Allow the gas to flow through the torch until you can touch it with your bare hands and *then* turn the air off.

17. Weld a "T" joint using the procedure described in Steps 10 through 14. See Fig. 15-14. Be sure to weld both sides of the joint.

142

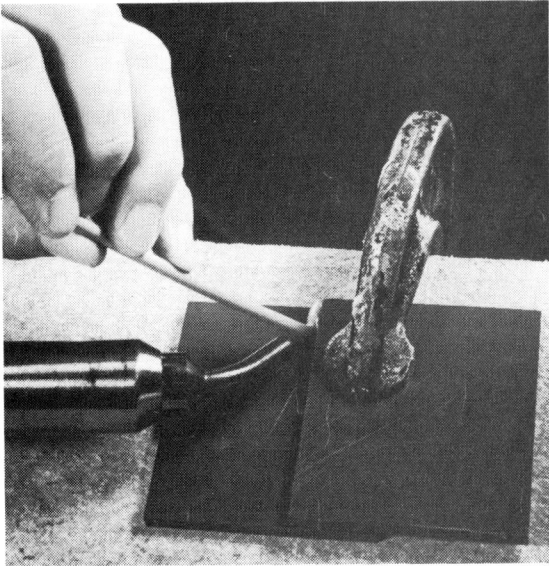

Fig. 15-10   Holding rod to root of weld.

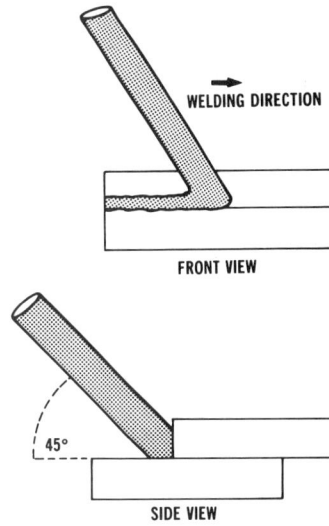

WELDING DIRECTION

FRONT VIEW

45°

SIDE VIEW

Fig. 15-11   Rod slanted back.

Fig. 15-12   Correctly held rod.

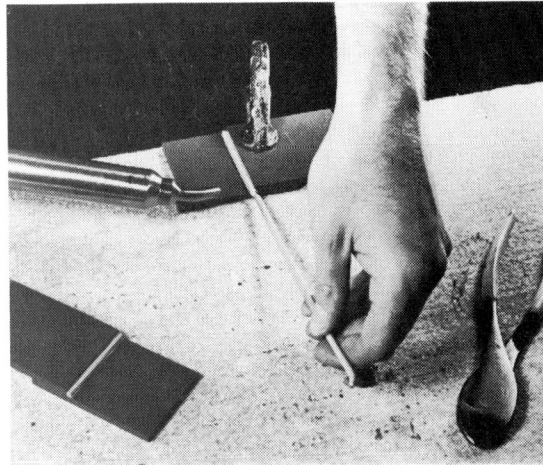

Fig. 15-13   Completed weld.

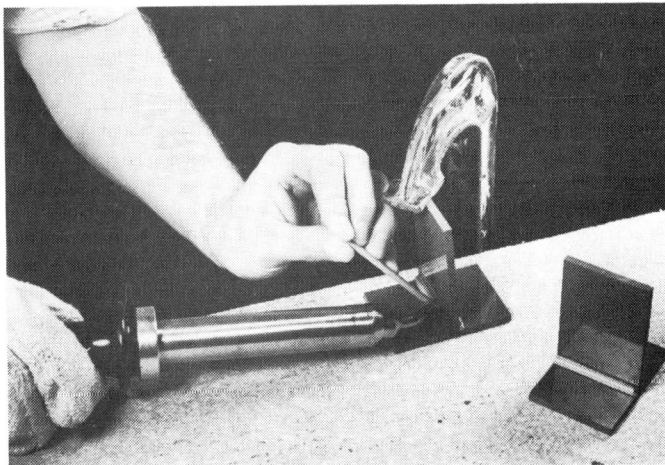

Fig. 15-14   Welding a "T" joint.

Hot Air Welding

18. Cut two equal size pieces of plastic to make a butt joint.

19. Sand a bevel on each piece to a 60° included angle. Be sure to leave a 1/32″ flat on the bottom of each beveled edge.

20. Clamp the plastic so that there is a 1/32 root gap (space) between each beveled edge.

21. Weld the butt joint using the procedure described in Steps 10 through 14. See Fig. 15-15.

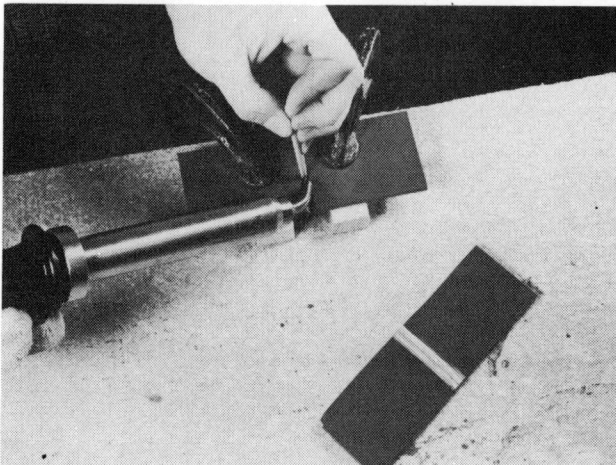

Fig. 15-15 Welding a butt joint.

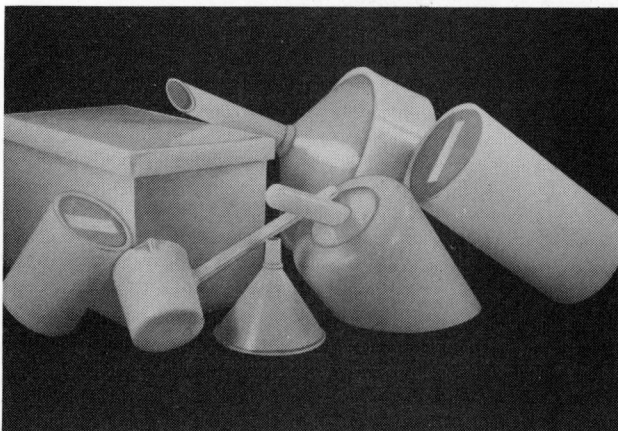

Fig. 15-16 Examples of hot air welded products.

CONCLUSION

Many hot air welded products are shown in Fig. 15-16. These products are made of polyethylene and polypropylene. They were welded with matching rods. The welding method shown in this assignment is called *hot air welding.* It is used for custom welding products. Hot air welding requires the use of both hands. This welding process is generally slow.

The welding speed can be increased many ways. One way is to use a speed welding tip. It can be used with most welding rods.

Another way to weld faster is with triangular welding rods. These rods are used in joints that need many round welding rod beads to fill. Generally, one pass with a triangular rod fills and finishes the joint.

Welding rods can be bought in many shapes, colors, sizes, and plastic types. Rods must be made of the same plastic as the plastic to be welded. Rod colors are often matched with the plastic being welded. Also, rod sizes vary and should be matched with the plastic thickness being welded. Always follow the manufacturer's

Section IV: Fabricating Plastics

recommendations for the proper rod size to use.

Rod shapes must be considered. A round rod can be used for many welding joints. Flat rods or strips are often used to weld butt joined plastic. Also, triangular rods are used to make deep V, corner, fillet, edge, and lap welds in one pass.

REVIEW

1. Name three hot air welded plastic products.

2. Explain two ways to increase hot air welding product production.

3. Explain where round, flat, and triangular welding rods are used.

4. Describe how hot air welding works.

5. Explain where and why an inert gas atmosphere is used for hot gas welding.

6. List four skills a good hot air welder must have.

7. Name the plastic(s) you welded.

8. Show how to make homemade welding rod.

9. Describe each step for hot air welding a PVC lap joint.

10. Describe how to increase and decrease the hot air welding temperature.

11. Show cold, hot, and very hot PVC welded joints.

12. Describe how to prepare and weld a butt joint.

13. Describe the torch tip fanning technique.

Hot Air Welding

14. Describe the welding rod starting position for a PVC lap joint.

SELECTED BIBLIOGRAPHY

Agranoff, Joan, ed. *Modern Plastics Encyclopedia. New York:* McGraw-Hill Book Company, 1976-77.

Baird, Ronald J. *Industrial Plastics.* South Holland, Illinois: The Goodheart-Willcox Company, Inc., 1971.

Hot Air Welding—Teacher's Manual. Indianapolis: Howard W. Sams & Company, Inc., 1974.

Kaminsky, S.J. and Williams, J.A. *Handbook for Welding and Fabricating Thermoplastic Materials.* Boston: Fidelity Press, 1964.

Patton, William J. *Plastics Technology: Theory, Design, and Manufacture.* Reston, Virginia: Reston Publishing Company Inc., 1976.

Richardson, Terry A. *Modern Industrial Plastics.* Indianapolis: Howard W. Sams & Company, Inc., 1974.

Rosato, Dominick V., ed. *Plastic Industry Safety Handbook.* Boston: Cahners Books, 1973.

EQUIPMENT AND MATERIAL SUPPLIERS

1. Ain Plastics Inc., 160 MacQuesten Parkway So., Mt. Vernon, New York 10550.

2. Brodhead-Garrett, 4560 East 71st Street, Cleveland, Ohio 44105.

3. Cadillac Plastics & Chemical Company, P.O. Box 810, Detroit, Michigan 48232.

Section IV: Fabricating Plastics

4. Cope Plastics, Inc., 4441 Industrial Drive, Godfrey, Illinois 62035.

5. Delvies Plastics, Inc., 2320 South West Temple, P.O. Box 1415, Salt Lake City, Utah 84110.

6. Graves-Humphreys, Inc., 1948 Franklin Road, P.O. Box 1347, Roanoke, Virginia 24033.

7. Industrial Arts Supply Company, 5724 West 36th Street, Minneapolis, Minnesota 55416.

8. Kamweld Products Company, Inc., 742 Providence Highway, P.O. Box 91, Norwood, Massachusetts 02062.

9. Laramy Products Company, Inc., Rt. 5N, Lyndonville, Vermont 05851.

10. Laurel Valley Manufacturing Company, 1601 Laurel Rd., Lindenwold, New Jersey 08021.

11. McKilligan Industrial Supply Corporation, 494 Chenango Street, Binghamton, New York 13901.

Welding equipment can be purchased from any of the suppliers listed above. Plastics for hot air welding are available from suppliers 2, 3, 4, 5, 7, and 10.

Hot Air Welding

Fig. 16-1 Impulse sealing machine.

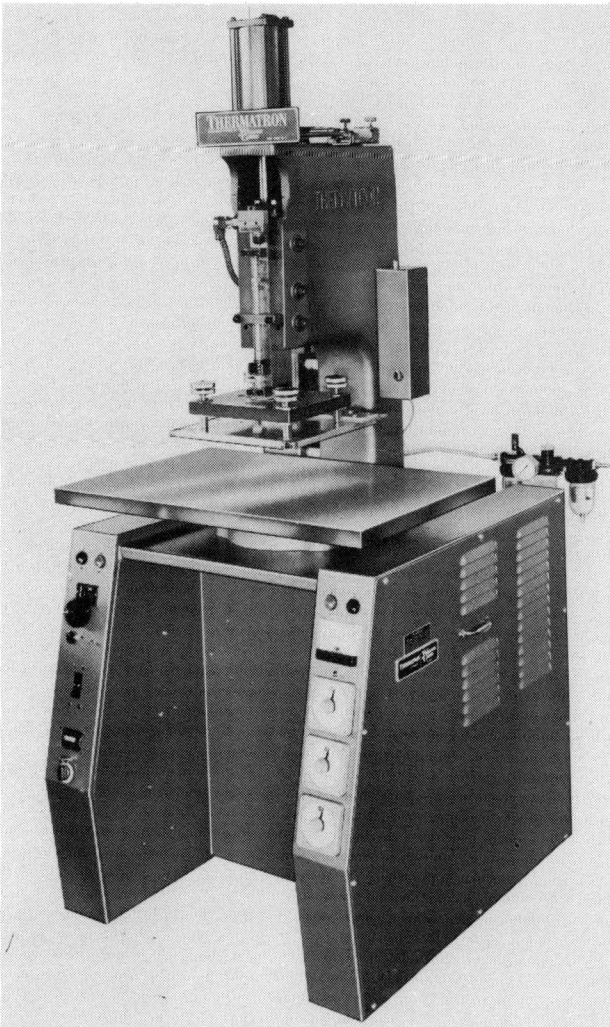

Fig. 16-2 Radio frequency sealing machine.
(Courtesy Thermotron Division/Solidyne, Inc.)

Assignment 16
Heat Sealing

OBJECTIVES

To prepare heat sealing equipment and plastic.

To use RF, impulse, and hand tool heat sealers to make heat sealed products.

INTRODUCTION

Heat sealing is a plastic bonding process. Thermoplastic under heat and pressure is bonded to itself or other materials. The bond is made when the heat-softened plastics are forced together. Heat sealing is used by the plastic packaging and fabricating industries. These industries make such products as packages, book covers, inflatable toys, pocket liners, flotation bags, rainwear, and tool holders.

One heat sealing process is *impulse sealing*. A demonstration of impulse sealing is shown in another assignment. Some impulse sealed plastics are nylon, vinyl, polyethylene, polypropylene, and fluorocarbon films. Shown in Fig. 16-1 is an impulse sealing machine with a 16″ sealing capacity.

Another heat sealing process is *radio frequency* (RF). It is also called high frequency or dielectric heat sealing. Some RF sealed plastics are polyvinyl chloride, ABS, nylon, polyether, polyester, cellulose acetate, and polyurethane foams, films, and fabrics. Shown in Fig. 16-2 is a radio frequency (RF) sealing machine. It has a 4 KW output and the frequency is stabilized.

In RF heat sealers, the sealer sends high frequency radio waves through the plastics. Some of these plastics do not conduct electricity (dielectrics) well. The radio frequency causes molecules of these plastics to vibrate. This action causes the plastic to heat and the plastic surfaces to soften. Then, pressure is applied to force the soft plastic together.

Hand tool heat sealing is similar to impulse sealing. With hand tool heat sealing, the heated jaw is always hot. The heated jaw is a heated bar, wheel, knife, or wide element. Generally, plastics sealed with an impulse sealer can also be hand tool heat sealed. A hand tool heat sealer is shown in Fig. 16-3. It has an electrically heated bar element clamp.

SAFETY

The following precautions should be taken when heat sealing:

1. Wear safety glasses.

2. Learn the safe operation of impulse, radio frequency (RF), and hand tool sealers.

3. Do not pinch your hands and fingers between the jaws or the jaw and the work table.

4. Understand the proper assembly, adjustment, and uses of heat sealing attachments.

5. Do not touch the hot sealing knife or sealing wheel.

6. Be sure the RF machine is shielded. This stops RF waves from interfering with radio, television, and other wireless communications.

EQUIPMENT AND MATERIALS

The equipment and materials needed for heat sealing are:

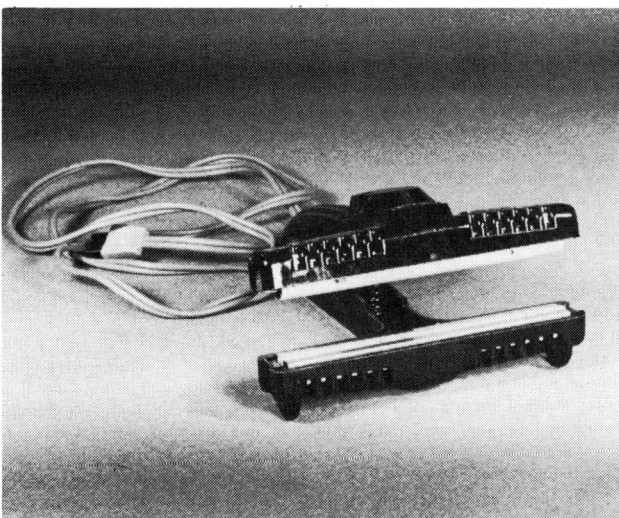

Fig. 16-3 Hand tool heat sealer.

Heat Sealing

Fig. 16-4 Impulse heat sealing equipment and materials.

Fig. 16-5 Connecting the footpedal.

1. Safety glasses.
2. Impulse sealer and footpedal.
3. Radio frequency (RF) sealer.
4. Hand tool sealer (bar element clamp).
5. Nylon, vinyl, polyethylene, polypropylene, or fluorocarbon film.
6. Scissors.

HEAT SEALING LAB PROCEDURES

The procedures for heat sealing plastics with an impulse sealer, radio frequency (RF) sealer, and hand tool sealer are described below. Obtain all the items listed in "Equipment and Materials" for each of the procedures.

Impulse Heat Sealing Procedure

1. Read Assignment 30 (Impulse Sealer). Steps 3 through 8 are a summary of the impulse heat sealing procedure. The equipment and materials used in this procedure are shown in Fig. 16-4.

2. Connect the footpedal unit to the impulse sealer. See Fig. 16-5.

3. Plug the impulse sealer into an outlet.

4. Set the sealing time with the timer dial. Your instructor will tell you the correct setting. Generally, the thicker the plastic, the longer the sealing time.

5. Hold the plastic between the upper arm and the sealing bar.

6. Press the footpedal fully until the buzzer sounds and then stops. See Fig. 16-6. Some machines have a light instead of a buzzer.

7. Keep the footpedal pressed down for the amount of time on the

timer. This lets the sealed plastic cool. A finished impulse sealed product is shown in Fig. 16-6.

8. Do Steps 5, 6, and 7, with each side of the product. The product (a pocket liner) will then be ready to use.

Radio Frequency (RF) Heat Sealing Procedure

1. Adjust the RF sealer power control switch. See Fig. 16-7. Your instructor will tell you the proper adjustment for Steps 1, 2, 3, and 4.

2. Set the RF sealer delay start timer. See Fig. 16-8. This is the period of time the upper jaw touches the plastic before the RF turns on.

3. Adjust the RF sealing timer. See Fig. 16-8. This is the amount of time the machine is RF heat sealing.

4. Set the RF sealer dwell timer. See Fig. 16-8. This is the amount of time the plastic is cooling in the RF sealer.

5. Turn on the RF machine electrical power switch. See Fig. 16-9.

6. Put the plastic parts to be sealed between the movable sealer jaw and the work table.

7. Turn on the RF switch.

8. Press both hand control switches at the same time. See Fig. 16-10. Be sure to use both hands. This lets the movable RF jaw move down. After the jaw touches the plastic and work table, the seal is made. The movable jaw will automatically open.

9. Remove the RF sealed product. See Fig. 16-11. This product is the front part of a bowling bag.

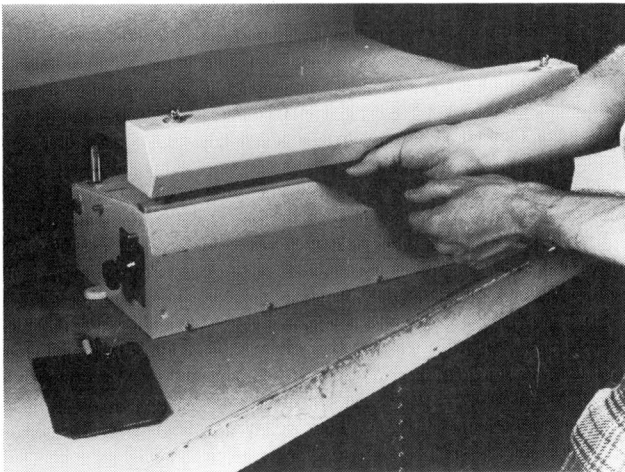

Fig. 16-6 Plastic inserted with footpedal pressed.

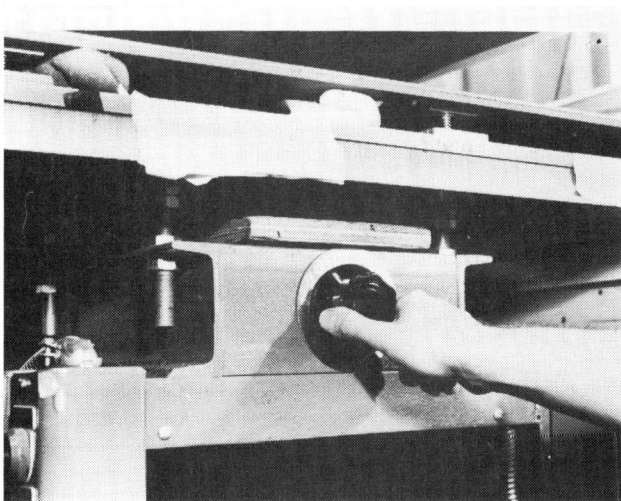

Fig. 16-7 Adjusting RF sealer power control switch.

Heat Sealing

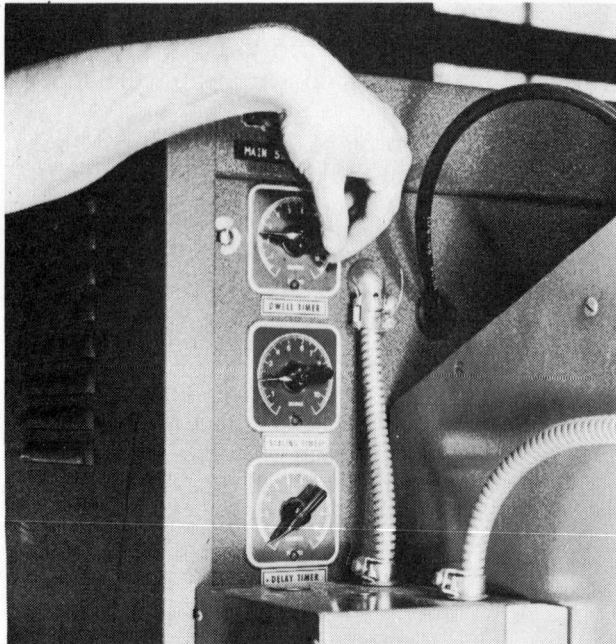

Fig. 16-8   Adjusting RF sealing timer.

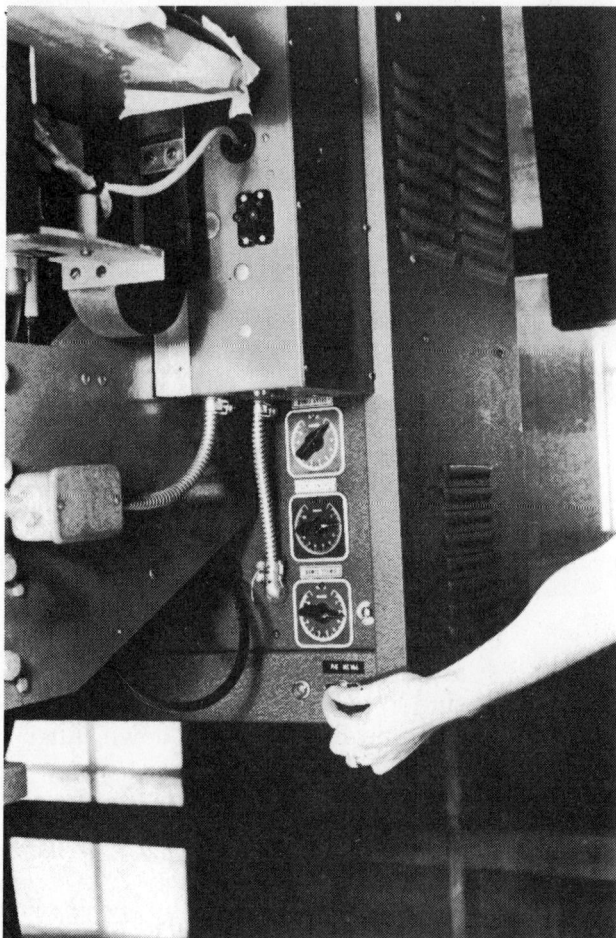

Fig. 16-9   Turning on power switch.

Hand Tool Heat Sealing Procedure

1. Plug the hand clamp sealer into an outlet. Let the heated jaw heat to the proper temperature.

2. Hold the overlapped pieces of plastic between the upper and lower jaws.

3. Press the upper jaw fully and snugly against the lower heated jaw. See Fig. 16-12. Leave the jaws clamped together the number of seconds needed to heat seal the plastic. Your instructor will tell you the amount of time to use. Generally, the thicker the plastic, the longer the sealing time. **Remember: once the sealer is plugged in, one jaw will heat to the sealing temperature of the plastic.** This jaw will stay hot until the sealer is unplugged.

4. Open the sealer jaw.

5. Remove the plastic. See Fig. 16-13. If plastic sticks to the jaws, use less sealing time. Also, the jaws can be cleaned by wiping them with a silicone paste release agent.

6. Repeat Steps, 2, 3, and 4 to make the other side of the product. The product (a pouch for papers) will then be ready to use.

CONCLUSION

Shown in Fig. 16-14 are many products that were impulse, RF, and hand tool sealed. All thermoplastics cannot be heat sealed with the same type of heat sealer. Be sure to use the correct heat sealer for the plastic being sealed. Some plastics cannot be heat sealed. They are often coated with a heat sealable plastic. This allows the plastic to be heat sealed.

Each heat sealing process shown

Section IV: Fabricating Plastics

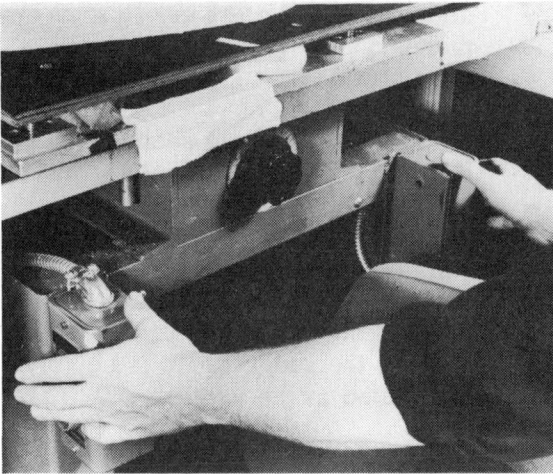

Fig. 16-10   Pressing hand control switches.

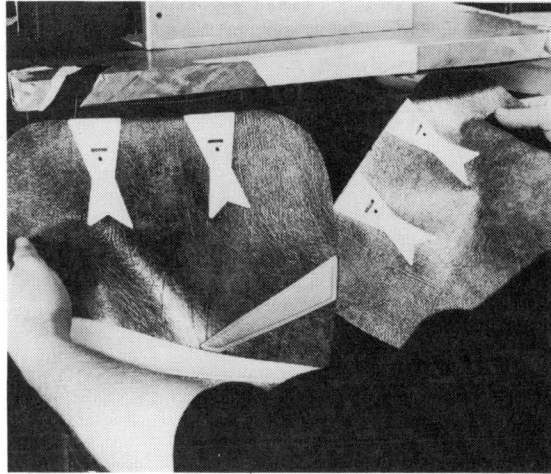

Fig. 16-11   Removing RF sealed product.

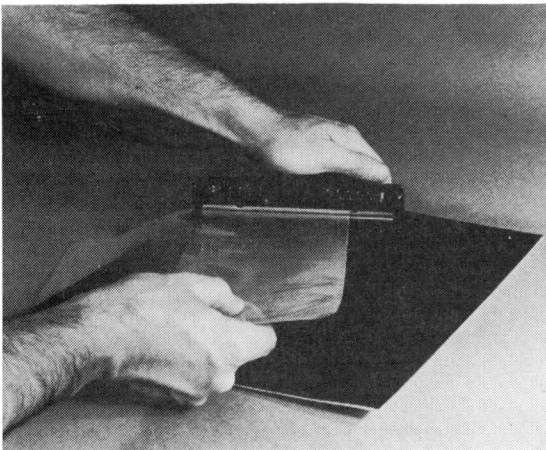

Fig. 16-12   Sealer jaws clamped together.

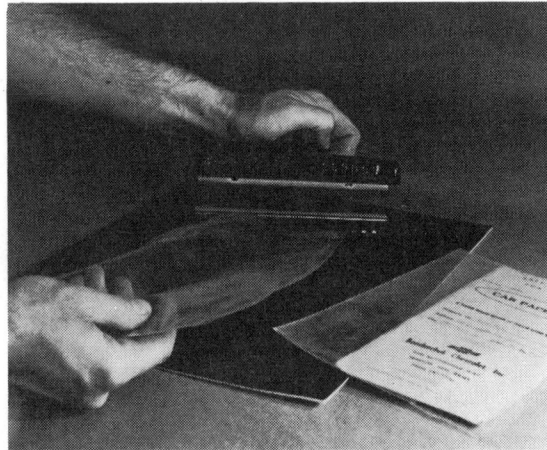

Fig. 16-13   Removing the plastic.

Fig. 16-14   Examples of impulse, RF, and hand tool sealed products.

Heat Sealing

has its own advantages. Impulse sealing lets the plastic quickly heat and cool under pressure. This makes a neat and wrinkle-free weld.

Radio frequency heat sealing has advantages. A high frequency weld can be made through a liquid. Different shaped welds (round, square, etc.) can be made. This is done by using different shaped electrodes. Also, in high frequency sealing, the sealing heat is equally distributed in the joint.

Hand tool heat sealing equipment is economical. Hand tool units seal overlapped plastic. See Fig. 16-15. The plastic is overlapped and placed on a clean, hard, wooden surface. A clean, heated wheel is pushed or pulled along the seam. Light pressure is used, and the seal is made.

Another hand tool heat sealer and cutter is the electric knife. See Fig. 16-16. The plastic is overlapped and placed on a clean, hard, wooden surface. A clean, hot knife edge is lightly pulled over the plastic. If a straight seal is wanted, a steel scale is used as a sealing edge guide.

There are many ways to heat seal plastics. Be sure to use the correct way to heat seal the plastic being used. This will give the best results.

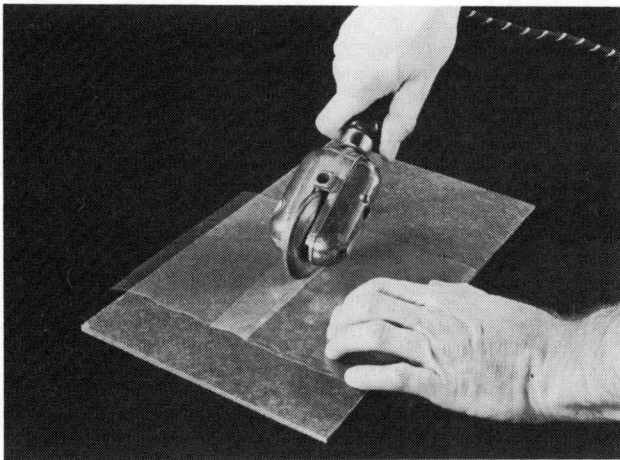

Fig. 16-15   Sealing overlapped plastic.

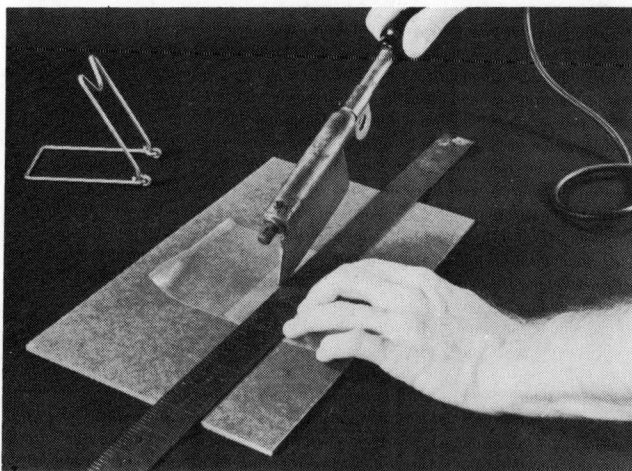

Fig. 16-16   Electric knife (hand tool heat sealer and cutter).

REVIEW

1. Name the plastic you impulse, RF, and hand tool heat sealed.

2. Describe each step in the RF heat sealing process.

3. Describe each step for hand tool heat sealing.

4. Describe how RF heat sealing works.

5. Name four products that are made by heat sealing.

6. Describe how hand tool heat sealing works.

7. Define *heat sealing.*

8. Name an advantage of impulse, RF, and hand tool heat sealing.

9. Explain why some plastics are laminated with a heat sealable film.

10. How can the plastic be stopped from sticking to the hand tool heat sealer?

SELECTED BIBLIOGRAPHY

Agranoff, Joan, ed. *Modern Plastics Encyclopedia.* New York: McGraw-Hill Book Company, 1976-77.

Baird, Ronald J. *Industrial Plastics.* South Holland, Illinois: The Goodheart-Willcox Company, Inc., 1971.

Briston, J.H. *Plastic Films.* New York: John Wiley and Sons, 1974.

Milby, Robert V. *Plastics Technology.* New York: McGraw-Hill Book Company, 1973.

Patton, William J. *Plastics Technology: Theory, Design, and Manufacture.* Reston, Virginia: Reston Publishing Company, Inc., 1976.

Richardson, Terry A. *Modern Industrial Plastics.* Indianapolis: Howard W. Sams & Company, Inc., 1974.

Rosato, Dominick V., ed. *Plastics Industry Safety Handbook.* Boston: Cahners Books, 1973.

EQUIPMENT AND MATERIAL SUPPLIERS

1. Ain Plastics, Inc., 160 MacQuesten Parkway South, Mt. Vernon, New York 10550.

2. Chemetron Corporation, Votator Division, 10300 Bunsen Way, Jeffersontown, Kentucky 40299.

3. Cope Plastics, Inc., 4441 Industrial Drive, Godfrey, Illinois 62035.

4. Frisch Division, Allen Industries, Inc., 1414 W. Wabansia Avenue, Chicago, Illinois 60622.

5. Industrial Arts Supply Company, 5724 West 36th Street, Minneapolis, Minnesota 55416.

6. Radio Frequency Company, 50 Park Street, Medfield, Massachusetts 02052.

7. Thermatron, Division of Solidyne Inc., 60 Spence Street, Bayshore, New York 11706.

8. Thermo-Dielectric Machine Company, 170 Tillary St., Brooklyn, New York 11201.

RF sealers can be purchased from suppliers 2, 4, 6, 7, and 8 listed above. Hand tool sealer can be purchased from suppliers 1, 3, and 5. Other heat sealing equipment and materials can be purchased from the suppliers listed in Assignment 30.

Section V
Laminating
Plastics

Assignment 17
High Pressure
Laminating

OBJECTIVES

To make a high pressure laminating "sandwich."

To adjust a laminating press.

To make a high pressure laminated product.

INTRODUCTION

High pressure laminating is an industrial process used to make rigid sheet material. Fig. 17-1 shows an industrial, high pressure laminating machine.

The laminating process works as follows. Sheets of wood, cloth, metal, glass, or paper are impregnated with thermosetting resins. The resins actually soak through the material. The resins used include phenolics, ureas, melamines, alkyds, silicones, epoxies, or polyesters. The impregnated sheets are then dried. This leaves the sheets well impregnated with dry resin.

The sheets are cut to size. They are then stacked and placed between metal plates. This is called a "sandwich." The "sandwich" is placed in a press. Press pressure and heat are now placed on the "sandwich." Press pressure varies from 1000 to 10,000 psi. The press heat ranges from 280° F to 400° F. Pressure and heat are kept on the "sandwich" for a period of time. This causes the resin to melt and bond the sheets together. The product is now one polymerized unit.

The press is cooled (optional) and the laminate removed. The laminate

Fig. 17-1 Industrial high pressure laminating machine. (Courtesy Reliable Rubber & Plastic Machinery Company)

is now cut to size and made into products. These products include table and counter tops, school desk tops, circuit boards, electrical insulator boards, motor mounting blocks, bearing retainers, kitchen utensil handles, tags, tubes, and rods.

A variety of high pressure laminates are available. Different laminates are made by using various combinations of resins and sheet materials. Various laminates are needed because many products require different chemical, electrical, and mechanical properties.

SAFETY

The following precautions should be taken when high pressure laminating:

1. Work in a well-ventilated area. Do not breathe plastic fumes because some are toxic.

2. Wear safety glasses and heat-resistant gloves.

3. Work on a heat-resistant surface.

4. Keep a general purpose fire extinguisher in the work area.

5. Learn the safe operation of the laminating press.

6. Do not touch the hot "sandwich" or hot press parts with your bare hands.

7. Be careful not to pinch your hands or fingers between press platens.

8. Be careful not to cut your hands or fingers on the "sandwich" plates.

9. Learn the safe operation of the high pressure laminating trimming equipment.

High Pressure Laminating

Fig. 17-2 School laboratory high pressure laminating press.

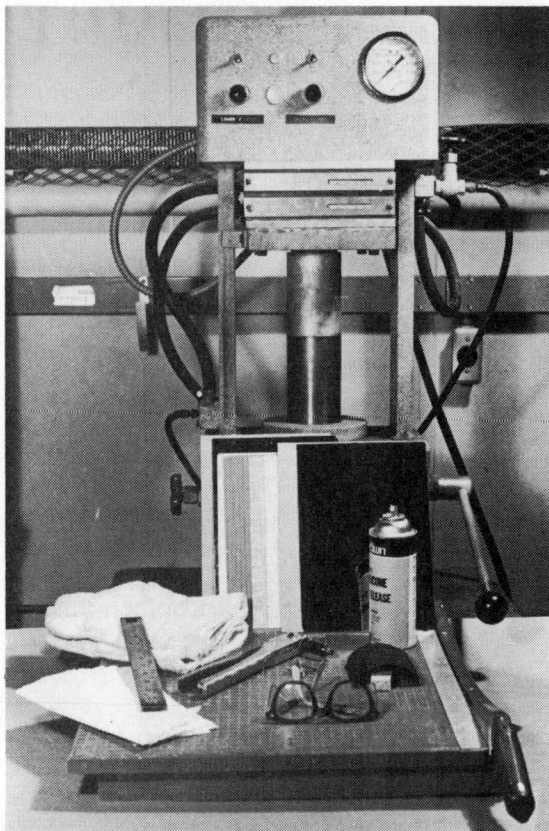

Fig. 17-3 High pressure laminating equipment and materials.

10. Keep your hands and fingers out of the high pressure laminating trimming cutter area.

EQUIPMENT AND MATERIALS

The equipment and materials needed for high pressure laminating are:

1. Safety glasses.
2. Heat-resistant gloves.
3. Heat-resistant surface.
4. Laminating press.
5. Two polished plates (stainless steel or chrome plated steel).
6. Two tag boards.
7. Release agent.
8. High pressure laminating core sheets.
9. High pressure laminating decorative sheets.
10. High pressure laminating overlay sheets.
11. Band saw or circular saw.
12. Paper cutter.
13. Pop rivet gun and pop rivets.
14. Sanding block and abrasive paper.
15. Ruler.
16. Clean cloth or paper towels.

HIGH PRESSURE LAMINATING LAB PROCEDURE

Shown in Fig. 17-2 is a school laboratory high pressure laminating press. Your school shop may have a similar unit. Be certain that you are familiar with all the controls on your shop press *before* using it. The procedure is as follows:

1. Obtain all the items listed under "Equipment and Materials" needed to high pressure laminate a project. See Fig. 17-3.

2. Pump the hydraulic press

Section V: Laminating Plastics

Fig. 17-4 Turning on the heater switches.

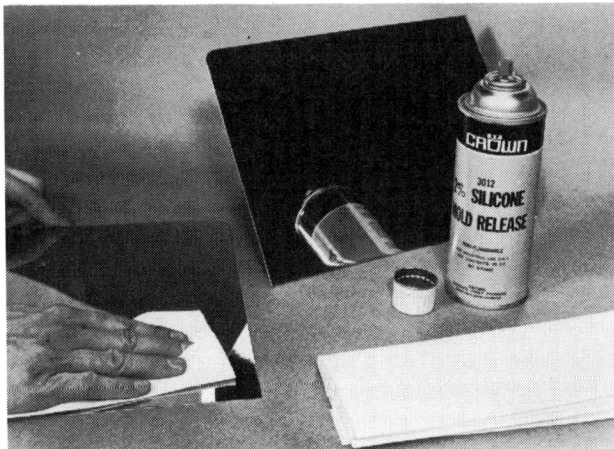

Fig. 17-5 Polishing laminating plates.

platens together until they just touch.

3. Turn on the heater switches for the upper and lower press platens. See Fig. 17-4. Platen heating is speeded up by heating them while they touch each other (kissing). Make sure the platen cooling water is turned off. Let the platens heat to the temperature recommended by the plastic manufacturer. The platens were heated to 300° F for this activity. Different high pressure laminating materials require different press temperatures. **Make sure that the instructor has set the platen heater thermostats.** They must be adjusted to the operating temperature of the plastics to be laminated. Keep each platen at its operating temperature through Step 16.

4. Spray each laminating plate with a release agent.

5. Polish each plate with a cloth or paper towel. See Fig. 17-5. The plates are made of chrome plated or stainless steel. Do not scratch the plates. Scratches on the plates will appear on the finished laminate surface. Make sure all dirt and grease have been removed from the plates.

6. Obtain two decorated sheets, two overlay sheets, and enough core sheets to produce a 3/16" thick finished product. See Fig. 17-6.

7. Cut each sheet ½" larger than the size of the product. See Fig. 17-7. Ask your instructor to give you the project measurements. A small clip board is made in this activity. Make sure each sheet is clean and free of grease.

8. Make up the high pressure laminating "sandwich" in the order listed below. See Fig. 17-8.

High Pressure Laminating

Fig. 17-6 "Sandwich" materials.

Fig. 17-7 Cutting sheets.

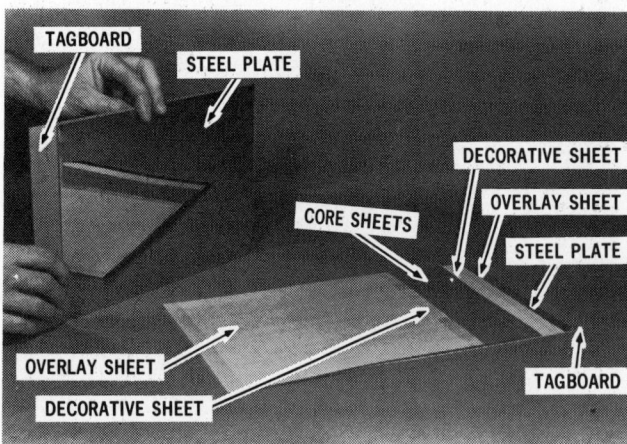

Fig. 17-8 Preparing laminating "sandwich."

a. tagboard
b. steel plate (stainless or chrome plate)
c. overlay sheet
d. decorative sheet
e. core sheets
f. decorative sheet
g. overlay sheet
h. steel plate (stainless or chrome plate
i. tagboard

9. Determine the pressure for the press by finding the *area* of one of the laminating sheets. Example: a 6″ × 8″ laminating sheet will have an area of 48 sq. in.

10. Multiply that area by the pressure (psi) your instructor gives you. The answer will be the press gauge reading. Example: 48 sq. in. × 1000 psi = 48,000 lbs. to be read on the pressure gauge.

11. Place the "sandwich" between each of the platens. See Fig. 17-9. Wear heat-resistant gloves to do this.

12. Pump the platens together and compress the "sandwich" until a slight amount of resistance is felt. See Fig. 17-10. A zero pressure reading should be read on the pressure gauge.

13. Open the platens about 1/16″ after 30 seconds. This is done by turning the platen release valve counter-clockwise. See Fig. 17-11. This is called the *breathing* or *degassing* step. Trapped gas can now escape.

14. Close the pattern release valve.

15. Pump the platens quickly together to the pressure calculated in Step 10. Read the pressure on the pressure gauge. See Fig. 17-12.

16. Keep the platens and "sandwich" under pressure for about fifteen

Section V: Laminating Plastics

Fig. 17-9 Placing "sandwich" between press platens.

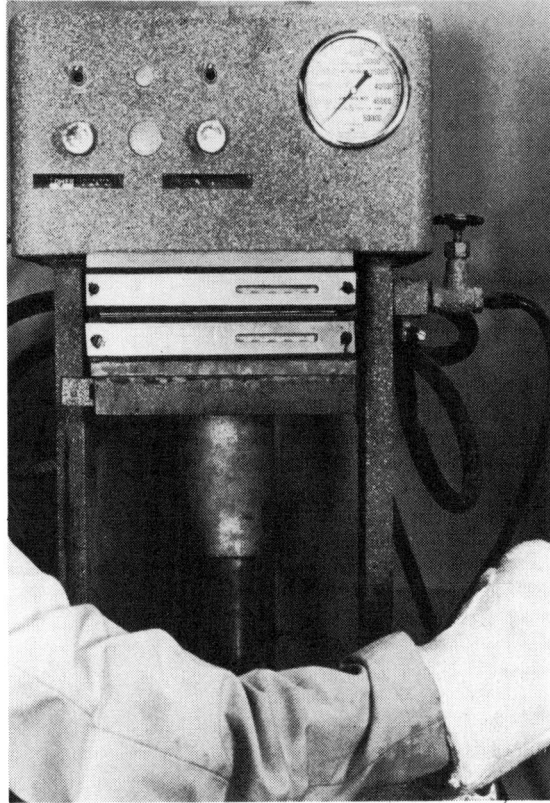

Fig. 17-10 Compressing the laminating "sandwich."

Fig. 17-11 Opening the platens.

Fig. 17-12 Pumping the platens together.

High Pressure Laminating

Fig. 17-13 Removing "sandwich" from press.

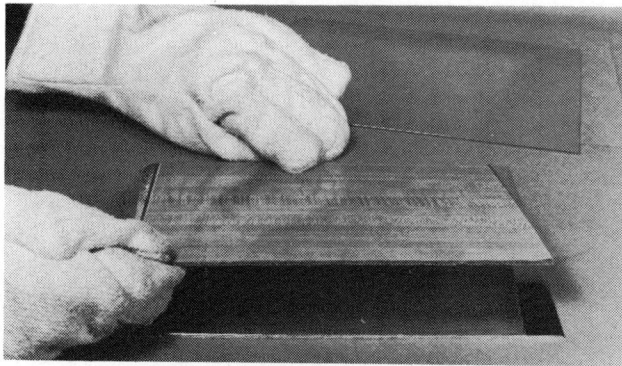

Fig. 17-14 Peeling laminate from metal plate.

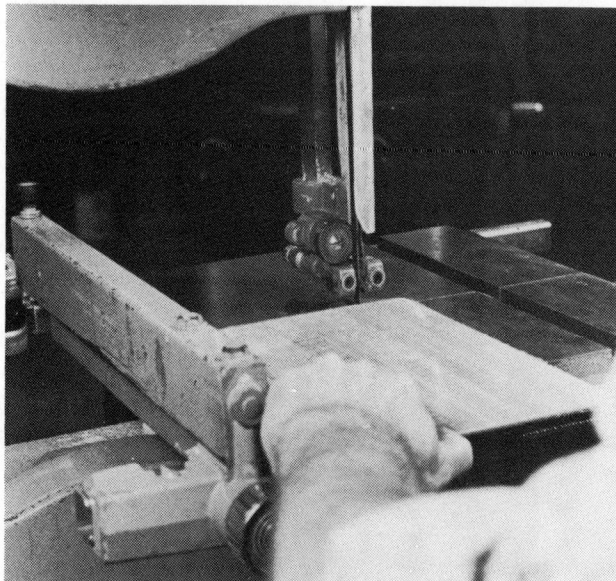

Fig. 17-15 Cutting laminate to size.

minutes. This is the high pressure laminating curing stage.

17. Turn off each platen heater switch.

18. Turn on each platen water cooling valve. Let the platen and "sandwich" cool under pressure until they reach about 140°F.

19. Turn off the water valves.

20. Open the platens about ½".

21. Remove the "sandwich" as in Fig. 17-13. Wear heat-resistant gloves to protect your hands from cuts and burns.

22. Separate the metal plates. Be sure to wear heat-resistant gloves.

23. Peel the laminate from the metal plate by hand. See Fig. 17-14. Do not touch the plates with a metal tool. A metal tool will scratch the plates. Dirt or scratches on the plates may make it hard to remove the laminate. If the laminate comes apart, there may not have been enough press pressure or heat.

24. Check the laminate for burns or surface marks. Burns may be caused by too much heat or too long a cure cycle. Marks may be caused by a dent or scratch on the metal plates.

25. Check the laminate surface for a white haze. This may be caused by not cooling the laminate. Using a light overlay sheet over a dark decorative sheet may also cause a white haze.

26. Cut the laminate to the finished size. Ask your instructor what sizes to use. See Fig. 17-15. Use a band saw with a metal cutting blade. A table saw with a fine-toothed tungsten carbide blade can be used. Follow your instructor's directions for

Section V: Laminating Plastics

Fig. 17-16 Sanding laminate edges.

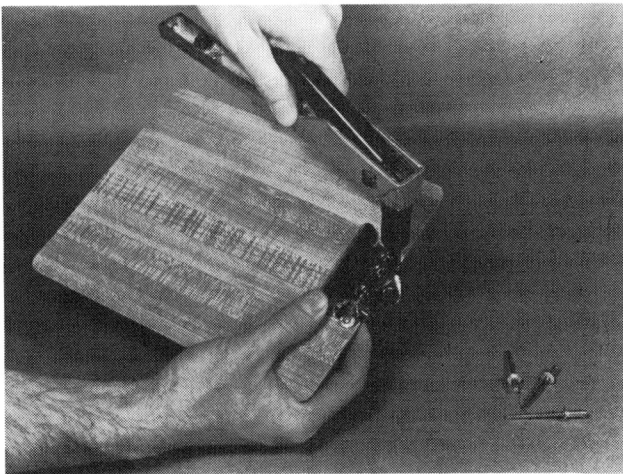

Fig. 17-17 Riveting clip to laminate.

Fig. 17-18 Examples of high pressure laminated products.

cutting your laminate with these machines.

27. A smooth finish can be made on the laminate edges by sanding them. See Fig. 17-16.

28. Locate two holes on the cut laminate. These are used to fasten the clip to the board. Your instructor will tell you where to locate the two holes.

29. Drill the two holes in the laminate. Your instructor will tell you what drill size and speed to use.

30. Pop rivet the clip to the laminate. See Fig. 17-17. The clip board is ready for use.

CONCLUSION

Many high pressure laminated products are shown in Fig. 17-18. Most of these products were laminated using the procedure just described. These products include various sizes of clip boards, a circuit board, spark plug lead puller, hot dish pad, luggage I.D. tag, key chain tags, playing chips, pencil holder, tool holder, knife handle, desk name signs, pen holder, cutting block, electrical insulator board, and gear paperweight.

The process just described is a slow one. Industrial high pressure laminating is much faster. To speed up laminate production, larger press platens are used. More products can be made from the larger platens. Also, presses are used with more than two platens. Finally, placing more than one "sandwich" between each pair of platens increases production.

High pressure laminating can be done continuously at high speed. The sheets are fed from rolls. They are first passed through a resin

High Pressure Laminating

impregnation tank and a resin drying oven. They are pressed under high pressure and heat between rollers or metal belts. The laminate is then trimmed and cut to shipping size.

REVIEW

1. Define *high pressure laminating.*

2. Describe each step of the high pressure laminating cycle.

3. Name the type of resin impregnated core sheet used in this activity.

4. List three products made by the high pressure laminating process.

5. Explain the reason for the degassing cycle.

6. Describe how the high production, high pressure laminating process works.

7. Name two ways to increase the output of the press used in this activity.

8. Describe how the press pressure was calculated for your product.

9. Describe how to stop a laminate from delaminating.

10. Name two high pressure laminate sheet materials.

11. Name two high pressure laminate resins.

SELECTED BIBLIOGRAPHY

Agranoff, Joan, ed. *Modern Plastics Encyclopedia.* New York: McGraw-Hill Book Company, 1976-77.

Baird, Ronald J. *Industrial Plastics.* South Holland, Illinois: The Goodheart-Willcox Company, Inc., 1971.

Formica Corporation. *Plastic Laminate Instruction Manual.* Cincinnati: Formica Corporation, 1971.

Low- and High-Pressure Laminating—Teacher's Manual. Indianapolis: Howard W. Sams & Company, Inc., 1974.

Milby, Robert V. *Plastics Technology.* New York: McGraw-Hill Book Company, 1973.

Richardson, Terry A. *Modern Industrial Plastics.* Indianapolis: Howard W. Sams & Company, Inc., 1974.

Rosato, Dominick V., ed. *Plastics Industry Safety Handbook.* Boston: Cahners Books, 1973.

Seymour, Raymond B. *Modern Plastics Technology.* Reston, Virginia: Reston Publishing Company, Inc., 1975.

EQUIPMENT AND MATERIAL SUPPLIERS

Press, plates, and tagboard may be purchased from any of the suppliers listed in Assignment 18 (Low Pressure Laminating). Resin impregnated core stock, resin impregnated overlay stock, and resin impregnated decorative stock may be purchased from any of the following suppliers:

1. Formica Corporation, Sub-American Cynamid Company, 120 E. Fourth Street, Cincinnati, Ohio 45202.

2. Ralph Wilson Plastics Company, 600 General Bruce Drive, Temple, Texas 76501.

3. Resopreg Products, Division of Pioneer Plastics, Pionite Road, Auburn, Maine 04210.

High Pressure Laminating

Assignment 18
Low Pressure Laminating

OBJECTIVES

To adjust a laminating press.

To make a low pressure laminating "sandwich."

To make a low pressure laminated product.

INTRODUCTION

Low pressure plastic laminates are products permanently sandwiched between two sheets of plastic. The finished product is called a "laminate." Low pressure laminates are made with pressures under 1000 psi.

Products are low pressure laminated to protect them from dirt and wear. Examples of products which may be low pressure laminated include business cards, photographs, menus, place mats, documents, awards, paperweights, desk sets, newspaper clippings, announcements, and game board covers.

High production low pressure laminating is often used by industry. It is also used in the laboratory when only a few items are being made. In this process, the product to be laminated is placed between two sheets of thermoplastic. This is called a "sandwich." The "sandwich" is then placed in a heated platen press under pressure. The pressure ranges from 200 to 900 psi. While the "sandwich" is in the press, the plastic softens. This softened plastic bonds to the product and to itself. The press is cooled to room temperature and the laminate removed. The laminate is then trimmed to its finished size.

SAFETY

The following precautions should be taken when low pressure laminating:

1. Work in a well-ventilated area. Do not breathe plastic fumes because some are toxic.

2. Work on a heat-resistant surface.

3. Wear safety glasses and heat-resistant gloves.

4. Keep a general purpose fire extinguisher in the work area.

5. Do not touch the hot "sandwich" or hot press parts with your bare hands.

6. Learn the safe operation of the laminating press.

7. Be careful not to pinch your hands or fingers between the press platens.

8. Be careful not to cut your fingers on the "sandwich" plates.

EQUIPMENT AND MATERIALS

The equipment and materials needed for low pressure laminating are:

1. Safety glasses.
2. Heat-resistant gloves.
3. Heat-resistant surface.
4. Laminating press.
5. Two polished plates (stainless steel or chrome plated steel).
6. Release agent.
7. Vinyl, acetate, acrylic, or polyester.
8. Two tagboards.
9. Article to be laminated.
10. Ruler.
11. Scissors or paper cutter.
12. Clean cloth or paper towels.

Low Pressure Laminating

LOW PRESSURE LAMINATING LAB PROCEDURE

1. Look at the low pressure laminating press. Make sure you know where the pressure, pressure release, heat, and cooling controls are located. See Fig. 18-1.

2. Obtain all the items listed under "Equipment and Materials" needed to low pressure laminate a product. See Fig. 18-2.

3. Pump the hydraulic press platens together until they just touch.

4. Turn on the heater switches for the upper and lower press platens. See Fig. 18-3. Platen heating is speeded up by heating them while they touch each other (kissing). Make sure the platen cooling water is turned off. Let the platens heat to the temperature recommended by the plastic manufacturer. (For example, vinyl has a heat range of 320°-350° F.) Make sure that your instructor has set the platen heater thermostats. They must be adjusted to the operating temperature of the plastics to be laminated. Be sure to keep each platen at its operating temperature through Step 19.

5. Lightly spray each laminating plate with a mold release agent.

6. Polish each plate with a cloth or paper towel. See Fig. 18-4. The plates are made of chrome plated or stainless steel. Do not scratch the plates. Scratches on the plates will appear on the finished laminate surface. Make sure all dirt and grease have been removed from the plates.

7. Obtain the product to be laminated (business cards, photograph, etc.).

Fig. 18-1 Low pressure laminating press.

Fig. 18-2 Low pressure laminating equipment and materials.

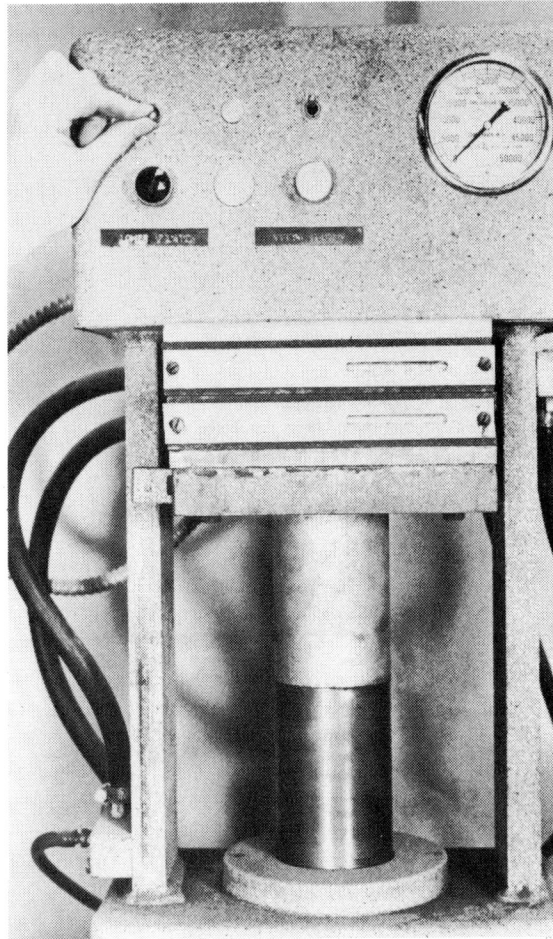

Fig. 18-3 Turning on heater switches.

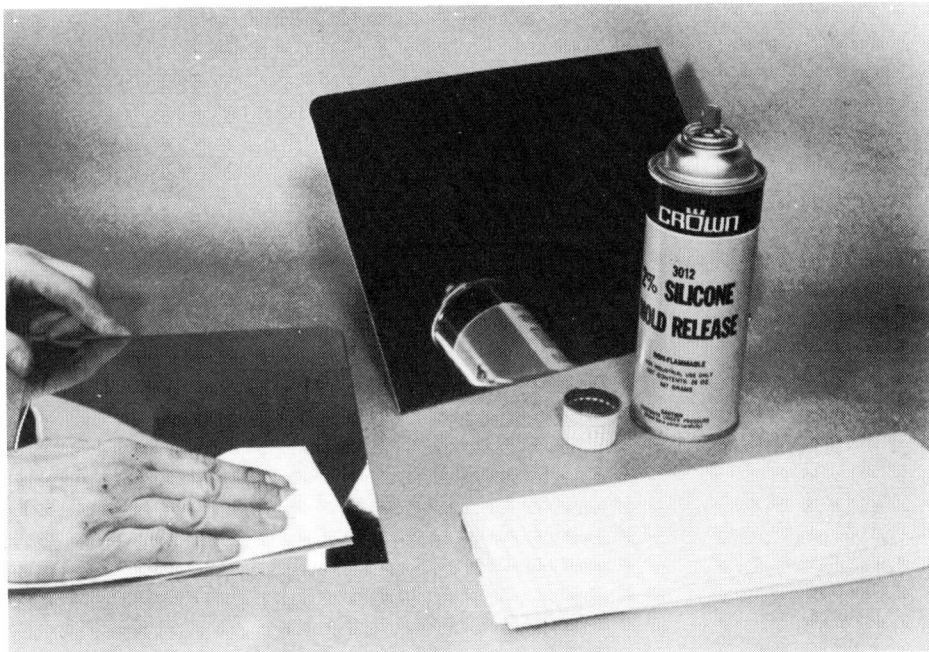

Fig. 18-4 Polishing laminating plates.

Low Pressure Laminating

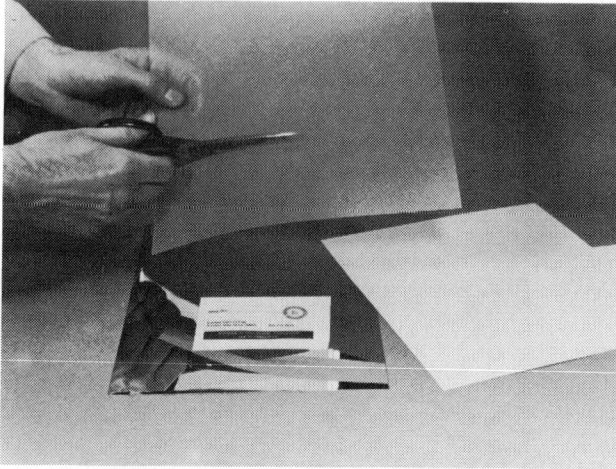

Fig. 18-5 Cutting plastic to size.

Fig. 18-6 Preparing laminating "sandwich."

8. Ask your instructor for the plastic to be laminated. Use the plastic your instructor recommends.

9. Cut the two pieces of plastic one inch larger than the product. This will make a one half inch border around it. See Fig. 18-5. Make sure the product and the plastic are clean and free of grease.

10. Make up the low pressure laminating "sandwich" in the order listed below. See Fig. 18-6.

 a. Tagboard (cardboard),
 b. Steel plate (stainless or chrome plated steel),
 c. Plastic,
 d. Product,
 e. Plastic,
 f. Steel plate (stainless or chrome plated steel),
 g. Tagboard (cardboard).

Be careful not to bump or jar the "sandwich" after it is put together. Bumping the "sandwich" may cause each part to move and separate.

11. Determine the proper pressure for the press. First, find the area of the item to be laminated. (Example: an item 3" x 4" will have an area of 12 sq. in.) If more than one item is being laminated, use the sum of all the areas.

12. Multiply the total area by the pressure (psi) given you by your instructor. The answer will be the **press pressure gauge setting.** (Example: 12 sq. in. x 300 psi = 3600 lbs. read on the press pressure gauge.) The low pressure laminating pressure of vinyl is 300 psi.

13. Place the "sandwich" between each press platen. See Fig. 18-7. Do not bump the "sandwich." *Wear heat-resistant gloves.*

14. Pump the platens together and compress the "sandwich" until a

Section V: Laminating Plastics

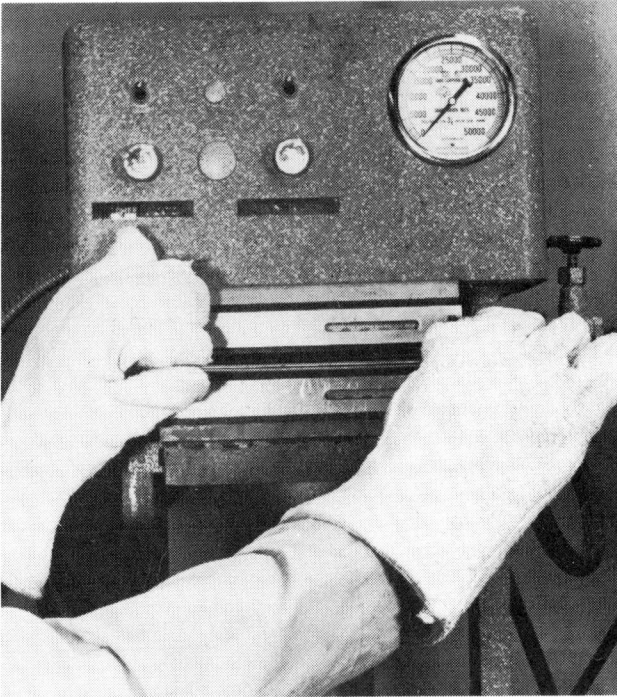

Fig. 18-7   Placing "sandwich" between press platens.

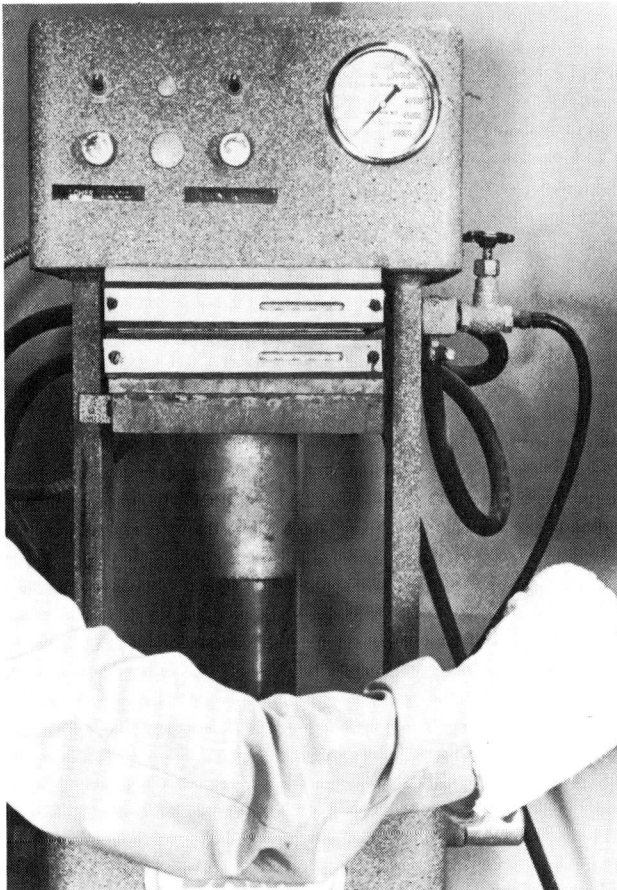

Fig. 18-8   Compressing the laminating "sandwich."

slight amount of resistance is felt. See Fig. 18-8. (Read **zero** pressure on the pressure gauge.)

15. Open the platens about ¹/₁₆" after 30 seconds. This is done by turning the platen release valve counterclockwise. See Fig. 18-9. This is the *breathing* or *degassing* step. Trapped gas can now escape. This step may be repeated if necessary.

16. Close the platen release valve.

17. Pump the platens quickly together to the pressure in Step 12. Read this pressure on the pressure gauge.

18. Keep the platens and "sandwich" under heat and pressure for 2 to 5 minutes. This is the low pressure laminating curing stage.

19. Turn off each platen heater switch.

20. Turn on each platen water cooling valve. See Fig. 18-10. Let the platen and "sandwich" cool under pressure until they reach room temperature.

21. Turn off the water valves.

22. Open the platens about ½".

23. Remove the "sandwich." See Fig. 18-11. Wear heat-resistant gloves to protect your hands from cuts and burns.

24. Separate the metal plates. Be sure to wear the heat-resistant gloves.

25. Peel the laminate from the metal plate with your fingernails. See Fig. 18-12. If it is hard to remove, check the plates for dirt or scratches. These may be the cause of the problem. **Do not touch the plates with a metal tool.** A metal tool will scratch the plates.

Low Pressure Laminating

Fig. 18-9 Opening the platens.

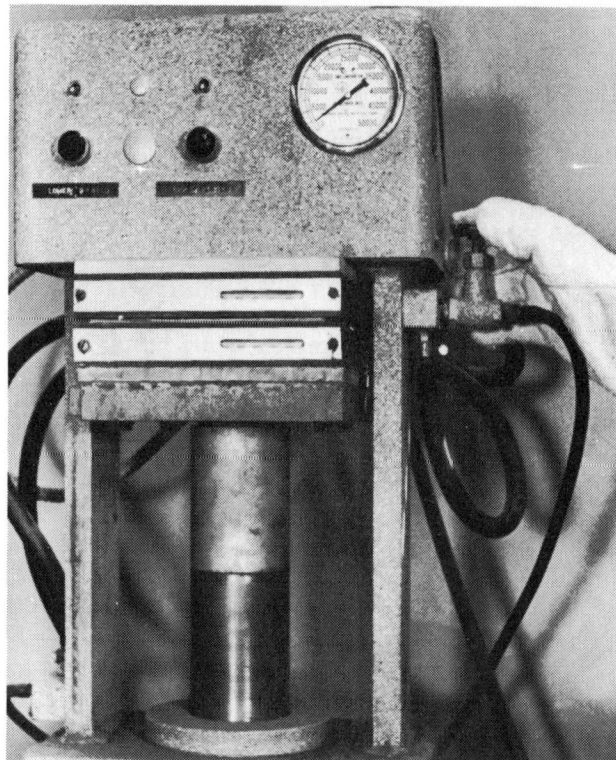

Fig. 18-10 Turning on platen water cooling valve.

26. Trim the laminates with scissors or a paper cutter. See Figs. 18-13. Leave ⅛" to ¼" plastic border around the laminated product. This prevents the laminate from coming apart or delaminating.

27. Check the laminate surface for marks. Some marks may be caused by a dent in the metal plates. Flow marks often result from overheating, low pressure, or a long curing cycle.

28. Check for tears, burns, or ink runs. These defects may be caused by overheating, a long curing cycle, or not enough "sandwich" preheat time.

29. Check for holes between the plastic and the product. This defect may be caused by moisture or grease in the "sandwich."

30. Check the laminate for spotty sealing. A low or unequally applied pressure may cause spotty sealing. Other causes include moisture or grease in the "sandwich" or dents on the metal plates.

CONCLUSION

Many low pressure laminated products are shown in Fig. 18-14. Most of these products were laminated using the procedure just described.

The postage stamp paperweight was made by placing stamps on one sheet of acrylic plastic. A second piece of acrylic plastic was then placed over the first sheet. The unit was then laminated using the procedure just described.

The desk set was made by laminating various colored acrylic plastics together. A name was engraved into one plastic sheet. This sheet was laminated and then the pen and letter holders were glued in place.

174

Fig. 18-11   Removing "sandwich" from press.

Fig. 18-12   Peeling laminate from metal plate.

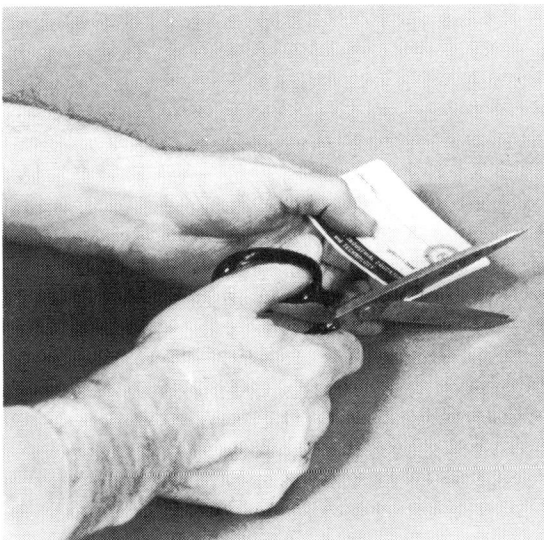
Fig. 18-13   Trimming the laminate.

Fig. 18-14   Examples of low pressure laminated products.

Low Pressure Laminating

The sketch was also low pressure laminated to the wooden plaque. The wooden plaque was made and placed in a "sandwich." The "sandwich" was built as follows:

a. Platen (heated),
b. Tagboard,
c. Steel plate,
d. Plastic,
e. Sketch,
f. Wood,
g. Platen (unheated).

After the "sandwich" was made, it was placed in the press under pressure. Only the top platen was heated during the lamination process. The "sandwich" was then laminated using the procedure just described.

The laminating process described here is slow. Mass production uses large press platens to make larger products. Laminating production can also be increased by using a press with more than two platens. Also, placing more than one "sandwich" between each pair of platens increases production.

Industry laminates products such as place mats with rollers. The plastic and the product are fed from rolls. They are pressed (laminated) as they pass through the heated rolls at high speed. The laminate is then cut to size and packaged.

REVIEW

1. Define *low pressure laminating.*

2. Describe each step of the low pressure laminating cycle.

3. Name the type of plastic used in this activity.

4. List three products produced by the low pressure laminating process.

5. Explain the purpose of the low

pressure laminating degassing cycle.

6. Describe how the high production, low pressure laminating process works.

7. Name two ways to increase the output of the press used in this activity.

8. Describe how the press pressure was calculated for your product.

9. Describe how a laminate can be stopped from delaminating.

SELECTED BIBLIOGRAPHY

Agranoff, Joan, ed. *Modern Plastics Encyclopedia.* New York: McGraw-Hill Book Company, Inc., 1976-77.

Baird, Ronald J. *Industrial Plastics.* South Holland, Illinois: The Goodheart-Willcox Company, Inc., 1971.

Low- and High-Pressure Laminating-Teacher's Manual. Indianapolis: Howard W. Sams & Company, Inc., 1974.

Richardson, Terry A. *Modern Industrial Plastics.* Indianapolis: Howard W. Sams & Company, Inc., 1974.

Rosato, Dominick V., ed. *Plastics Industry Safety Handbook.* Boston: Cahners Books, 1973.

Warner Electric Company. *How to Do Plastic Laminating for Pleasure and Profit.* Chicago: Warner Electric Company, Inc., 1976.

EQUIPMENT AND MATERIAL SUPPLIERS

1. Brodhead-Garrett, 4560 East 71st Street, Cleveland, Ohio 44105.

2. Cope Plastics, Inc., 4441 Industrial Drive, Godfrey, Illinois 62035.

3. Delvie's Plastics, Inc., 2320 South West Temple, P.O. Box 1415, Salt Lake City, Utah 84110.

4. Graves-Humphreys, Inc., 1948 Franklin Road, P.O. Box 1347, Roanoke, Virginia 24033.

5. Industrial Arts Supply Company, 5724 W. 36th St., Minneapolis, Minnesota 55408.

6. McKilligan Industrial Supply Corporation, 494 Chenango Street, Binghamton, New York 13901.

7. Paxton/Patterson, 5719 West 65th Street, Chicago, Illinois 60638.

8. Vicor Plastic Equipment, Inc., 231 E 1st Avenue, Roselle, New Jersey 07203.

9. Warner Electric Company, 1512 W. Jarvis Avenue, Chicago, Illinois 60626.

The plastic, press, plates and tagboard for use in this assignment may be purchased from any of the suppliers listed above.

Section VI
Polyester
Molding

Assignment 19
Matched Mold Hand Layup

OBJECTIVES

To prepare a matched mold.

To select and cut reinforcing material.

To make a matched mold hand layup product.

INTRODUCTION

Matched mold hand layup is a basic reinforced plastic molding process. It is used to make strong, flexible, and attractive products. The products are resistant to heat, cold, corrosion, water, and electric current.

A two part mold is used for this molding process. It may be made of wood, aluminum, steel, cement, or reinforced plastic. The mold surfaces should be sealed with an epoxy or urethane finish. All open mold surfaces receiving the layup are washed. After washing, the surfaces are coated with a release agent.

A clear or colored gel coat is catalyzed and put on the mold surfaces. The gel coat is then dried. Resin is catalyzed and put over the gel coat.

A layer of decorative material is put on each resin coated mold half. This is often an extra step. Then, more catalyzed resin and a layer of reinforcing material are put on the decorative material of each mold. These two materials are applied one after the other. This is done until the product is the right thickness.

The mold halves with the wet resin and material are put together. Light

180

pressure is put on the mold halves. While the mold is under pressure, the resin cures.

It cures into a hard material that makes the reinforcing material stick together. After the resin cures, the mold is opened. The product is removed and trimmed.

Shown in Fig. 19-1 are workmen doing a reinforced plastic hand layup. The workmen are rolling resin through fiber glass mat. They are also rolling air out of the resin and fiber glass mixture.

Many products are made with this method. A few of these products are clipboards, skateboards, trays, chairs, skis, and tool boxes.

SAFETY

The following precautions should be taken when hand layup molding:

1. Work in a well-ventilated area.

2. Wear a respirator when machine abrading (sanding) reinforced plastic.

Fig. 19-1 Reinforced plastic layup process. (Courtesy Owens-Corning Fiberglas Corporation)

Matched Mold Hand Layup

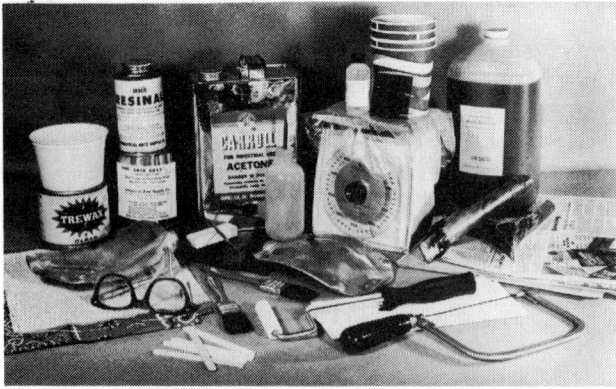

Fig. 19-2 Matched mold hand layup equipment and materials.

Fig. 19-3 Washing mold surfaces.

Fig. 19-4 Applying water soluble release agent.

3. Wear safety glasses and disposable gloves.

4. Keep a general purpose fire extinguisher in the work area.

5. Keep the resin away from heat and flame.

6. Do not mix catalyst and accelerators directly because it can cause a violent chemical explosion.

7. Use all hand layup resin parts as recommended by the manufacturer.

8. Protect your eyes and skin from contact with the resin or reinforcing material. If contact is made, *immediately* wash your skin and flush your eyes with clean, cold water. Rewash the skin with warm water to remove all trace of reinforcing material. Then, *get medical attention.*

EQUIPMENT AND MATERIALS

The equipment and materials needed for matched mold hand layup are:

1. Safety glasses.
2. Disposable gloves.
3. Matched mold.
4. Mold release agent (hard paste wax or water soluble).
5. Resin and catalyst.
6. Coloring agent (optional).
7. Gel coat.
8. Two pieces of fiber glass cloth reinforcing material (8″ × 10″).
9. Decorative cloth (2 pieces 8″ × 10″).
10. 16-oz. unwaxed paper cups.
11. Water and container.
12. Mixing sticks.
13. Acetone and container.
14. Squeegee.
15. Roller.

Section VI: Polyester Molding

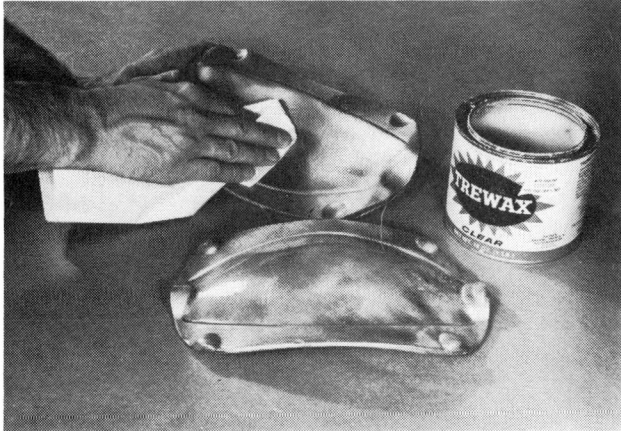

Fig. 19-5 Applying mold release wax.

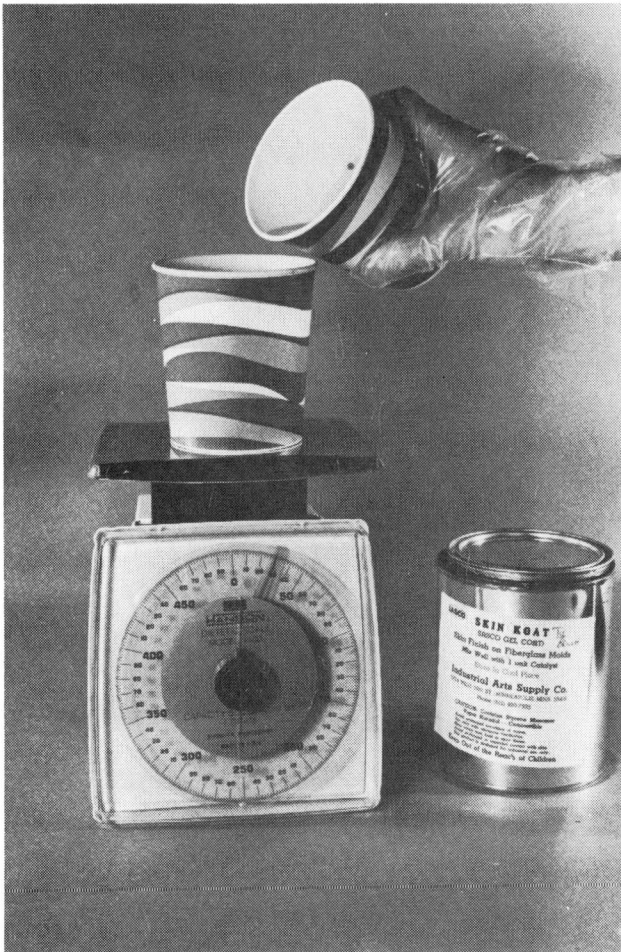

Fig. 19-6 Weighing gel coat.

16. Coarse, medium, and fine wet/dry sandpaper.
17. Sanding block.
18. 2 one-inch brushes.
19. Artist's brush.
20. 2 clamps.
21. Woodworking bench vise.
22. Coping saw.
23. Band saw.
24. Weight scales.
25. Dropper bottle.
26. Clean cloth or paper towels.
27. Newspapers.

MATCH MOLD HAND LAYUP LAB PROCEDURE

1. Obtain all the items listed under "Equipment and Materials" needed to match mold hand layup. See Fig. 19-2. Make sure you cover the bench top with newspapers or other throw away materials.

2. Wash the mold surfaces with water. See Fig. 19-3. Make sure all old resin, dirt, and release agents are removed. If a solvent is used, make sure it does not remove the mold finish. The mold shown is for making a small candy dish.

3. Brush each mold half layup surface with an even, bubble-free coat of water soluble release agent. See Fig. 19-4. Be sure to clean the brush in water.

4. Place three coats of mold release wax on the layup surface of each mold half. See Fig. 19-5. Buff each dry coat before applying the next coat.

5. Let the release agent dry on each mold half surface.

6. Weigh enough gel coat to cover both mold half layup surfaces. Fig. 19-6. Ask your instructor how much gel coat to weigh out.

7. Add the proper number of drops of catalyst to the gel coat. Ask your

Matched Mold Hand Layup

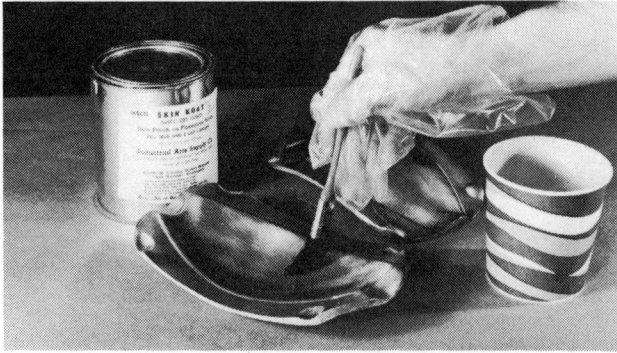

Fig. 19-7 Applying catalyzed gel coat to mold surface.

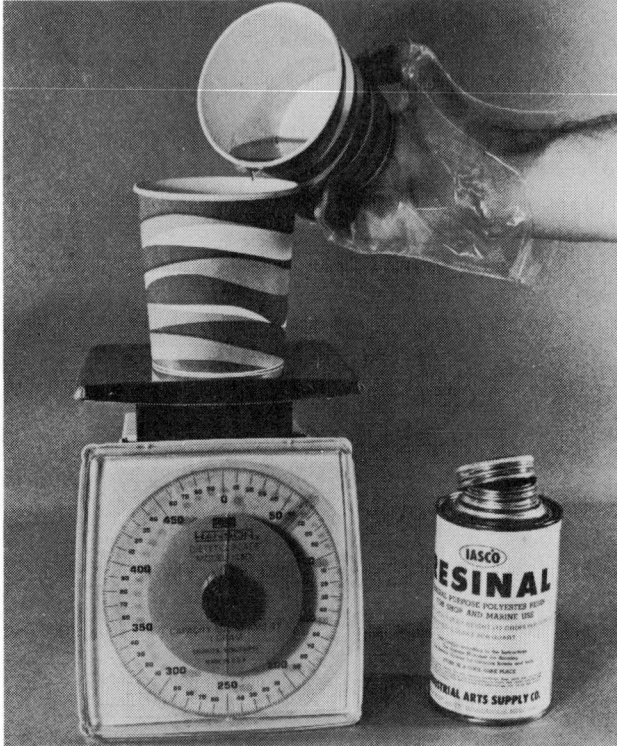

Fig. 19-8 Weighing resin.

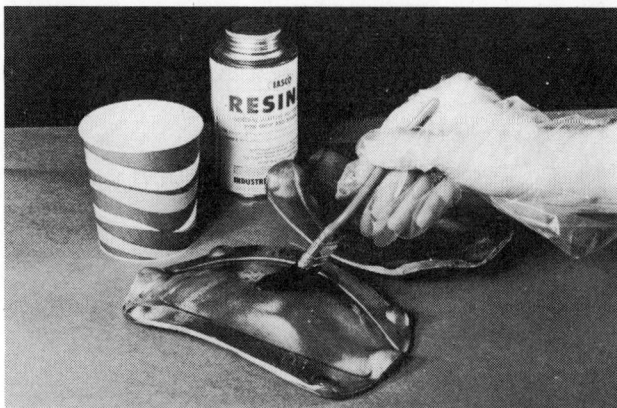

Fig. 19-9 Applying catalyzed resin to gel coated surface.

instructor how many drops to use. (Generally, up to 1% catalyst by weight is mixed with the gel coat.)

8. Mix the catalyst and gel coat well. Follow your instructor's mixing directions. Be sure to wear disposable gloves when handling gel coat, resin, and catalyst. Work in a well-ventilated room.

9. Brush a coat of gel coat on each mold half layup surface. See Fig. 19-7. The gel coat makes a waterproof outside finish on the product. Be sure to clean the brush in acetone immediately.

10. Let the gel coat dry.

11. Weigh enough resin to coat each mold half surface a few times. See Fig. 19-8. Your instructor will tell you how much to weigh out. Weigh out no more resin than can be used in ten minutes.

12. Add the proper number of catalyst drops to the resin. Again, ask your instructor about the correct number of drops. Generally, up to 1% catalyst by weight is mixed with resin.

13. Mix the catalyst and resin as shown in Step 8.

14. Brush a coating of catalyzed resin on the gel coat surface of each mold half. See Fig. 19-9.

15. Lay the decorative cloth on the wet resin surface of each mold half. See Fig. 19-10.

16. Brush a coat of catalyzed resin on the decorative cloth of each mold half. Be sure to apply the resin with a dabbing motion. This helps the resin to soak through the cloth (wetout). It also helps remove trapped air from between the cloth and mold surface.

17. Lay a piece of fiber glass cloth

Fig. 19-10 Placing decorative cloth on wet resin surface.

Fig. 19-11 Placing fiber glass cloth over wet resin surface.

Fig. 19-12 Wetting out the fiber glass cloth.

on the wet resin surface of each mold half. See Fig. 19-11.

18. Wet out the cloth as shown in Step 16. See Fig. 19-12. Catalyzed resin can also be worked through the cloth with a roller or squeegee.

19. Fit the mold halves together with the wet resin surfaces touching.

20. Place newspapers on the floor below the vise. Newspapers can also be placed in the vise. This protects the vise and floor from resin.

21. Put the mold into a woodworking bench vise. Be sure to place light pressure on the mold. Also, a few hand clamps may be placed around the unclamped parts of the mold.

22. Let the product cure overnight.

23. Remove the mold halves by hand. See Fig. 19-13. Do not pry the mold halves open with sharp tools. They will scratch the mold surfaces.

24. Trim the extra material from the product. Use a fine-toothed coping saw or a band saw with a metal cutting blade.

25. Sand the edges for a smooth finish. See Fig. 19-14. Start sanding with 100 grit, then 220 grit, and finish with 400 grit. Be sure to use a sanding block with the abrasive paper. This helps prevent the plastic edge from being rounded off.

26. Weigh, catalyze, and mix ½ oz. of gel coat. Refer to Steps 6, 7, and 8.

27. Brush a coat of catalyzed gel coat around the edge of the product. This adds a finish to the product edge. It also helps stop *delamination* (layers of cloth peeling apart).

Matched Mold Hand Layup

Fig. 19-13 Removing mold halves.

Fig. 19-14 Sanding the edges.

28. Look at the finished candy dish shown in Fig. 19-15. Four round legs can be W.E.P. or polyester cast. See Assignment 21 or 22 for the casting procedures. The legs can be attached to the bottom of the dish by resin or glue.

CONCLUSION

Fig. 19-16 shows many layup reinforced plastic products. The process just described uses woven fiber glass as the reinforcing material. It also uses polyester resin as the plastic material. Fiber glass is the most common product reinforcing material. It may be bought as a mat (chopped glass strands stuck together in a swirl pattern), woven roving (woven coarse fibers), fabmat (combination of mat on one side and woven fabric on the other side), and continuous roving (fiber glass strands wound together). Other materials are also used as reinforcing. A few of these include paper, asbestos mats, plastic fibers, graphite, burlap, and boron fibers.

The thermosetting resin used most for hand layup reinforced plastic is polyester. Thermosetting resins other than polyester are also used. A few of these include epoxy, phenolic, silicone, and melamine.

During hand layup, chemical curing takes place. A simple cure cycle is as follows. When the gel coat or resin is catalyzed, the catalyst breaks down. This causes the chemical release of items called "free radicals." They are attracted to certain chemical reactive areas of the resin molecules or chains. Now certain areas of each resin molecule or chain are reactive. They join (crosslink) to other reactive resin molecules or chains. This forms cross chains or bonds between the original resin chains. Heat (exothermic) is generated and given

Section VI: Polyester Molding

off by the product. After the reaction (cross-linking) stabilizes, the product cools. The original and new cross-linked chains (molecules) form tightly around the reinforcing. This makes the material stiff.

REVIEW

1. List three products made by the two piece mold hand layup process.

2. Describe each step in two part mold hand layup.

3. List two resins and three reinforcing materials used for making hand layups.

4. Name the material your mold was made of.

5. What is a gel coat?

6. Describe how polyester resin, glass reinforced products cure.

7. Name the mold release you used on your mold.

8. Describe the catalyst-to-resin ratio and the catalyst-to-gel coat ratio you used.

9. Explain how to wetout reinforcing material.

10. How can product delamination be stopped?

SELECTED BIBLIOGRAPHY

Agranoff, Joan, ed. *Modern Plastics Encyclopedia.* New York: McGraw-Hill Book Company, 1976-77.

Bacon, Clare E. *Fiberglas/Plastic Applications in Appliances and Equipment.* Toledo, Ohio: Owens-Corning Fiberglas Corporation, 1972.

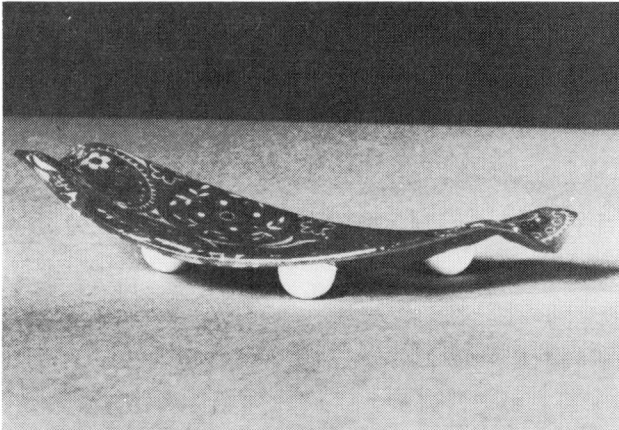

Fig. 19-15 Finished candy dish.

Fig. 19-16 Examples of hand layup reinforced plastic products.

Matched Mold Hand Layup

Baird, Ronald J. *Industrial Plastics. South Holland,* Illinois: The Goodheart-Willcox Company, Inc., 1971.

Milby, Robert V. *Plastics Technology.* New York: McGraw-Hill Book Company, 1973.

Owens-Corning Fiberglas Corporation. *Reinforced Plastics.* Publication No. 5-PL-3101-A. Toledo, Ohio: Owens-Corning Fiberglas Corporation, 1967.

Patton, William J. *Plastics Technology: Theory, Design, and Manufacture.* Reston, Virginia: Reston Publishing Company, Inc. 1976.

Reinforced Plastics—Teacher's Manual. Indianapolis: Howard W. Sams & Company, Inc., 1974.

Richardson, Terry A. *Modern Industrial Plastics.* Indianapolis: Howard W. Sams & Company, Inc., 1974.

Rosato, Dominick V., ed. *Plastic Industry Safety Handbook.* Boston: Cahners Books, 1973.

EQUIPMENT AND MATERIAL SUPPLIERS

1. Brodhead-Garrett, 4560 East 71st Street, Cleveland, Ohio 44105.

2. Cope Plastics, Inc., 4441 Industrial Drive, Godfrey, Illinois 62035.

3. Delvies Plastics, Inc., 2320 South West Temple, P.O. Box 1415, Salt Lake City, Utah 84110.

4. Industrial Arts Supply Company, 5724 West 36th Street, Minneapolis, Minnesota 55416.

Section VI: Polyester Molding

5. McKilligan Industrial Supply Corporation, 494 Chenango Street, Binghamton, New York 13901.

The following products may be purchased from any of the suppliers listed above: resin, color, reinforcement, molds, release agent, and gel coat.

Matched Mold Hand Layup

Assignment 20
Spray Layup

OBJECTIVES

To prepare a spray layup mold.

To properly adjust reinforcing material chopping and resin spraying equipment.

To make a spray layup product.

INTRODUCTION

Spray layup, using an open mold, is a simple reinforced plastic molding method. This process is often used to make large products. It is also used when a large number of products must be made. Spray layup is often used to make complex product shapes because the chopped fibers flow easily into the mold contours.

The spray layup process works as follows. A proper mold is first selected. These molds are often made of wood, aluminum, steel, cement, or reinforced plastic. The mold surfaces should be sealed with epoxy or a urethane finish. The mold surface is then cleaned and coated with a release agent. A gel coat is catalyzed and sprayed on the mold surface. The gel coat is dried. Catalyzed resin and chopped reinforcement are placed on the gel coat. If the materials are applied together, a combination chopper and spray gun is used.

Resin and chopped fibers are applied until the product is the right thickness. The resin and reinforcement are then hand rolled. This removes trapped air and the resin wets the reinforcement. It also makes the product dense. The product is left to cure. Then, the resin polymerizes into a finished product. After curing, the product is removed from the mold and trimmed.

Many products are made by spray layup. They include restaurant seats, boat hulls, free-form chairs, utility buckets, dome structures, sinks, small vehicle bodies, toys, industrial exhaust hoods, and pool equipment. Industrial spray layup work is shown in Fig. 20-1.

SAFETY

The following precautions should be taken when spray layup molding:

1. Work in a well-ventilated area.

2. Wear a respirator when machine sanding reinforced plastic.

3. Wear a respirator when applying resin and reinforcing material with the spray and chopper units.

4. Wear safety glasses and disposable gloves.

Fig. 20-1 Industrial spray layup operation. (Courtesy Owens-Corning Fiberglas Corporation)

Spray Layup

5. Keep a general purpose fire extinguisher in the work area.

6. Keep the resin away from heat and flame.

7. Do not mix catalyst and accelerators directly because it can cause a violent chemical explosion.

8. Use all spray layup resin parts as recommended by the manufacturer.

9. Learn the safe operation of the reinforcing material chopper unit.

10. Do not adjust the chopper while it is running.

11. Do not put your hands and fingers near the chopper cutter.

12. Clean (purge) the spray unit right after using it.

13. Protect your eyes and skin from contact with the resin or reinforcing material. If contact is made, *immediately* wash your skin and flush your eyes with clean, cold water. Rewash the skin with warm water to remove all trace of reinforcing material. Then, *get medical attention.*

EQUIPMENT AND MATERIALS

The equipment and materials needed for spray layup are:

1. Safety glasses.
2. Disposable gloves.
3. Spray unit.
4. Spray nozzle lid.
5. Resin spray container (unwaxed paper cups).
6. Spray booth.
7. Air line and air regulator.
8. Chopper unit.
9. Mold.
10. Mold release agent (hard paste wax or water soluble).

Fig. 20-2 Spray layup outfit.

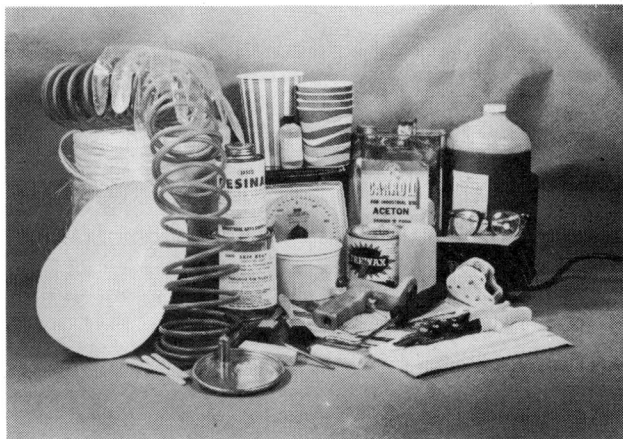

Fig. 20-3 Spray layup equipment and materials.

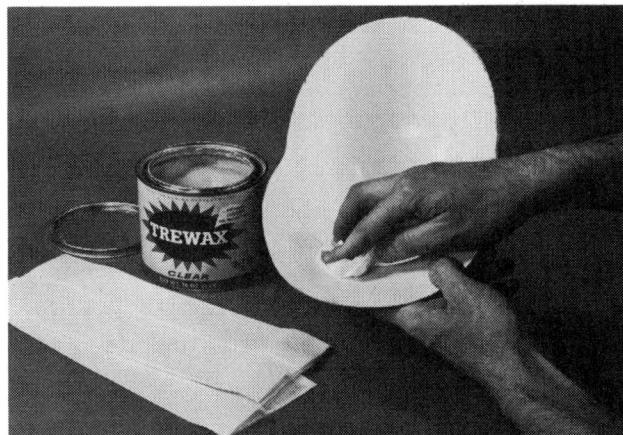

Fig. 20-4 Applying mold release wax.

11. Resin and catalyst.
12. Coloring agent (optional).
13. Gel coat.
14. Reinforcing material (roving).
15. 4 16-oz. unwaxed paper cups.
16. Water and container.
17. Mixing sticks.
18. Acetone and container.
19. Squeegee.
20. Roller.
21. Coarse, medium, and fine wet/dry sandpaper.
22. Sanding block.
23. 2 one-inch brushes.
24. Artist's brush.
25. Metal cutting shears.
26. Weight scales.
27. Dropper bottle.
28. Clean cloth or paper towels.
29. Newspapers.

SPRAY LAYUP LAB PROCEDURE

An inexpensive spray layup outfit is shown in Fig. 20-2. Polyester resin and gel coats are put on the mold with the spray unit. Reinforcing material is cut to size with the chopper unit. This is probably similar to the equipment you will be using in your school shop.

1. Obtain all the items listed under "Equipment and Materials" needed to spray layup. See Fig. 20-3.

2. Wash the mold surface with water. Make sure all old resin, dirt, and release agents are removed. If a solvent is used, make sure it does not remove the mold finish. The mold shown is for making a small helmet.

3. Apply three coats of mold release wax to the mold layup surface. See Fig. 20-4. Buff each dry coat before applying the next coat.

4. Brush the layup surface with an even, bubble-free coat of water-soluble release agent. See

Spray Layup

Fig. 20-5. Be sure to clean the brush in water.

5. Let the release agent dry.

6. Weigh enough gel coat to cover the mold surface. See Fig. 20-6. Ask your instructor how much to weigh out.

7. Add the proper number of drops of catalyst to the gel coat. Again, ask your instructor how many drops to use. Generally, up to 1% catalyst by weight is mixed with gel coat.

8. Mix the catalyst and gel coat well. Follow your instructor's mixing directions. Be sure to wear disposable gloves when handling gel coat, resin, and catalyst. Work in a well ventilated area.

9. Select the proper lid and nozzle unit (7/32" to ¼" nozzle opening).

10. Snap the lid and nozzle unit on to the top of the cup. See Fig. 20-7.

11. Place the cup and lid/nozzle unit onto the spray gun. See Fig. 20-8. The gun spring clip lock will hold the unit in place.

12. Adjust the air line pressure to about 30 psi.

13. Spray the gel coat on a scrap of plastic or wood. This will help you to see if the gun is spraying properly. Be sure to follow your instructor's recommendations for spray gun adjustments.

14. Place the mold in the spray booth.

15. Spray an even gel coat (1/32" thick) on the mold surface. See Fig. 20-9. Be sure to soak and clean the lid/nozzle unit in solvent (acetone or MEK) right after spraying. A small pipe cleaner or artist brush can be used to clean the nozzle opening.

Fig. 20-5 Applying water-soluble release agent.

Section VI: Polyester Molding

Fig. 20-6   Weighing the gel coat.

Fig. 20-7   Snapping lid and nozzle unit onto cup top.

Fig. 20-8   Placing lid and nozzle unit in spray gun.

Fig. 20-9   Spraying gel coat on mold surface.

Spray Layup

Throw the cup away when you are finished.

16. Let the gel coat cure.

17. Weigh enough resin to coat the mold surface. Ask your instructor how much to weigh out. Weigh out no more resin than can be used in ten minutes.

18. Add the proper number of drops of catalyst to the resin. Ask your instructor how many drops to use.

19. Mix the catalyst and resin.

20. Spray a coat of catalyzed resin on the gel coat surface. See Fig. 20-10.

21. Plug in the chopper.

22. Switch on the chopper cutters.

23. Thread roving into the cutter guard hole. See Fig. 20-11.

24. Run the chopper cutters at a medium speed.

25. Direct an even stream of chopped fiber glass roving over the resin-coated mold. See Fig. 20-12. Ask your instructor how *thick* to make the fiber glass-layer. Watch for the colored strand of fiber glass, the *tracer*. The amount of tracer in any one spot will tell you the thickness of the fiber glass layer.

26. Spray a coat of catalyzed resin over the layer of fiber glass. See Fig. 20-13.

27. Brush the resin with a dabbing motion through the fiber glass (wetout). See Fig. 20-14. Be sure all air bubbles are removed from the layer of resin-coated fiber glass. Catalyzed resin can also be worked through the fiber glass with a roller or a squeegee.

28. Keep applying and wetting out

Fig. 20-10 Spraying catalyzed resin.

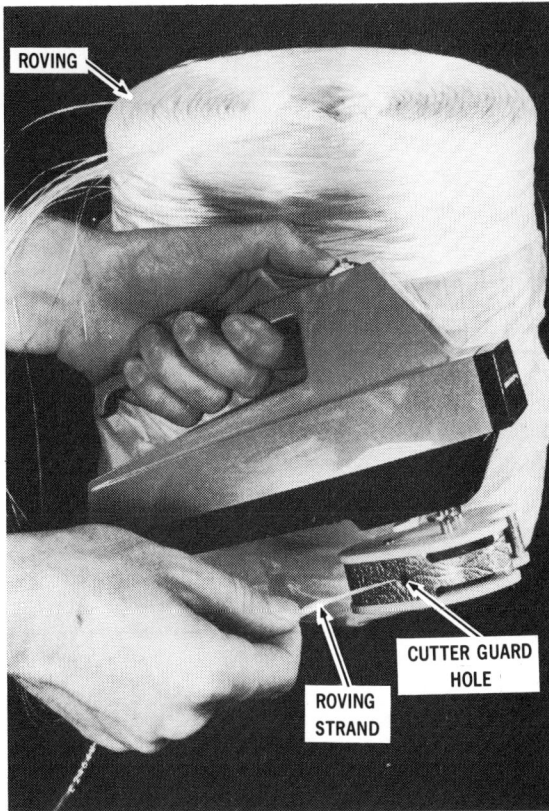

Fig. 20-11 Threading roving into cutter guard hole.

Fig. 20-12 Directing roving over resin coated mold.

Fig. 20-13 Spraying catalyzed resin coat.

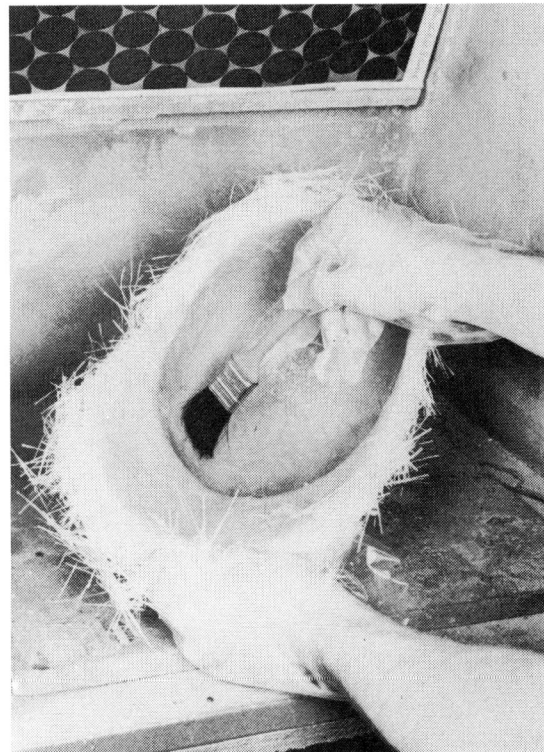

Fig. 20-14 Brushing resin through fiber glass.

197

Spray Layup

Fig. 20-15 Removing product from mold.

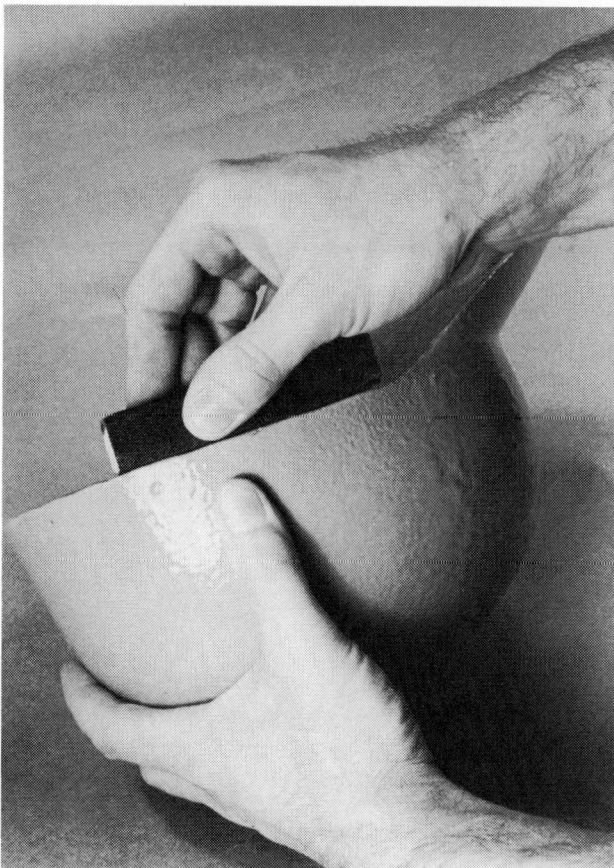

Fig. 20-16 Sanding the edges.

alternate layers of fiber glass and resin as shown in Steps 25, 26, and 29. Ask your instructor how many resin-coated fiber glass layers you should have. Be sure to immediately clean the brush, lid/nozzle, squeegee, and roller in solvent (acetone or MEK).

30. Let the product air-cure overnight.

31. Remove the product from the mold by hand. See Fig. 20-15. Do not remove the product from the mold with sharp tools. This action will scratch the mold.

32. Trim the extra fiber glass from the product with metal cutting shears.

33. Sand the product edges for a smooth finish. See Fig. 20-16. Start with 100 grit wet/dry, then use 220 grit, and finish with 400 grit. **Be sure to use a sanding block.** This helps prevent the plastic edge from being rounded off.

34. Weigh, catalyze, and mix ½ oz. of gel coat as shown in Steps 6, 7, and 8.

35. Brush a coat of catalyzed gel coat around the edge of the product. This adds a finish to the product edge. It also helps stop the product edge from delaminating (layers of cloth pealing apart). The finished helmet is shown in Fig. 20-17. It can be used when a helmet liner and head band are added.

CONCLUSION

Many spray layup reinforced plastic products are shown in Fig. 20-18. The process just described used fiber glass roving for the reinforcing material. *Roving* is strands of fiber glass wound together on a spool. It is bought by the weight per spool and by the number of monofilaments (ends) wound together per strand.

A commonly used thermoset resin for spray layup is polyester. Plastics other than polyesters also are now used. One type is epoxy. It is about 30% lighter than most polyester resins.

After the spray layup is finished, it is cured. The curing cycle can often be increased by slightly heating the product. This also speeds the product production rate.

During the molding of the part, a chemical curing process takes place. This process is explained in Assignment 19.

REVIEW

1. List three products made by the spray layup process.

2. Describe each step in the spray layup process.

3. Name two resins and one reinforcing material used for spray layups.

4. Name the material from which your mold was made.

5. Describe how polyester-resin glass reinforced products cure.

6. Name the mold release you used on your mold.

7. Describe the catalyst-to-resin ratio and the catalyst-to-gel coat ratio you used.

8. Describe how to densify, remove air, and wetout reinforcing material.

9. Explain why a combination chopper and spray unit is used for spray layup.

10. Give a reason for using a

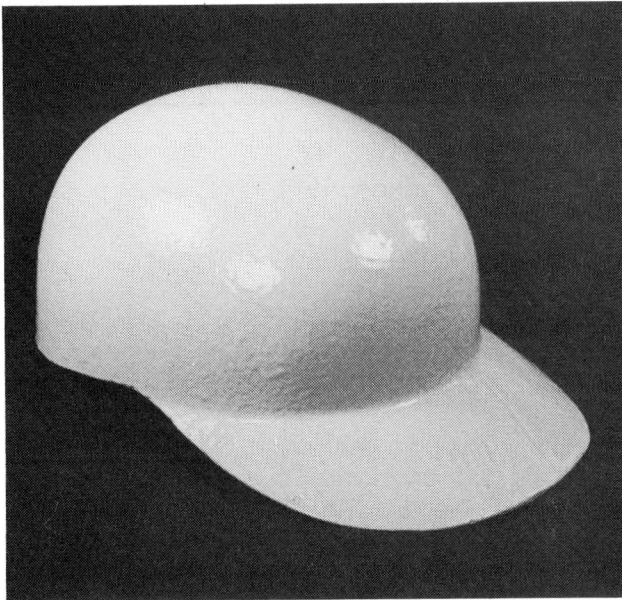

Fig. 20-17 Finished helmet.

Fig. 20-18 Examples of spray layup plastic reinforced products.

199

Spray Layup

separate chopper unit and spray unit for spray layups.

SELECTED BIBLIOGRAPHY

Agranoff, Joan ed. *Modern Plastics Encyclopedia.* New York: McGraw-Hill Book Company, 1976-77.

Bacon, Clare E. *Fiberglas/Plastic Applications in Appliances and Equipment.* Toledo, Ohio: Owens-Corning Corporation, 1972.

Baird, Ronald J. *Industrial Plastics.* South Holland, Illinois: The Goodheart-Willcox Company, Inc., 1971.

Milby, Robert V. *Plastics Technology.* New York: McGraw-Hill Book Company, 1973.

Owens-Corning Fiberglas Corporation. *Reinforced Plastics.* Publication No. 5-PL-3101-A. Toledo, Ohio: Owens-Corning Fiberglas Corporation, 1967.

Patton, William J. *Plastics Technology: Theory, Design, and Manufacture.* Reston, Virginia: Reston Publishing Company, Inc., 1976.

Reinforced Plastics-Teacher's Manual. Indianapolis: Howard W. Sams & Company, Inc., 1974.

Richardson, Terry A. *Modern Industrial Plastics.* Indianapolis: Howard W. Sams Company, Inc., 1974.

Rosato, Dominick V., ed. *Plastics Industry Safety Handbook.* Boston: Cahners Books, 1973.

EQUIPMENT AND MATERIAL SUPPLIERS

1. Brodhead-Garrett, 4560 East 71st Street, Cleveland, Ohio 44105.

Section VI: Polyester Molding

2. Cope Plastics, Inc., 4441 Industrial Drive, Godfrey, Illinois 62035.

3. Delvies Plastics, Inc., 2320 South West Temple, P.O. Box 1415, Salt Lake City, Utah 84110.

4. Industrial Arts Supply Company, 5724 West 36th Street, Minneapolis, Minnesota 55416.

5. McKilligan Supply Corporation, 494 Chenango Street, Binghamton, New York 13901.

Spray layup molds, resin, colorant, gel coat, and release agent can be purchased from any of the above listed suppliers. Spray guns are available from suppliers 1, 4, and 5. Particle choppers and roving can be purchased from suppliers 1, 3, 4, and 5.

Assignment 21
Clear Casting and Encapsulating

OBJECTIVES

To prepare a mold and sample.

To make a clear casting encapsulating the sample.

INTRODUCTION

Pouring a catalyzed liquid plastic into an open mold is called "clear casting." Encapsulating is the same technique except the plastic is cast around a sample in a mold. Such items as school emblems, family crests, jewelry, art objects, and giftware are made by the casting process. Items such as coins, leaves, advertisements, photographs, animal specimens, shells, flowers, and electrical components are encapsultated.

Clear casting and encapsulating work as follows. A mold is selected. The surfaces are cleaned and a release agent applied. A sample is selected, cleaned, and dried. Damp samples must be dehydrated (dried) before they are embedded in plastic. Sharp edges of certain samples (minerals) are dulled. This prevents cracks from forming in the cured casting.

The resin is then catalyzed. It is poured into the mold to a depth of ⅛". When the layer begins to gel, a second batch of resin is catalyzed. It is brushed on the sample. The sample is now placed on the first gel layer. Then, the second layer is poured around it. After this layer gels, another batch of resin is catalyzed and poured.

This procedure is repeated until the mold is filled. After the mold is filled, a glass cover or lid is placed on the mold. The casting is cured into a hard plastic. Then, the casting is removed from the mold, trimmed, and polished.

During casting, chemical curing takes place. A simple cure cycle is as follows. When the resin is catalyzed, the catalyst breaks down. This causes the chemical release of items called "free radicals." They are attracted to certain chemical reactive areas of the resin molecules or chains.

At this point, certain areas of each resin molecule or chain are reactive. They join (crosslink) to other reactive resin molecules or chains. This forms cross chains or bonds between the original resin chains. Heat (exothermic) is generated and given off from the product. After the reaction (crosslinking) stabilizes, the product cools.

SAFETY

The following precautions should be taken when clear casting and encapsulating:

1. Work in a well-ventilated area.

2. Wear safety glasses.

3. Wear disposable gloves.

4. Keep a general purpose fire extinguished in the work area.

5. Keep resin away from heat and flame.

6. Be sure to measure the catalyst correctly before you add it to the resin.

7. Protect your eyes and skin from contact with the resin. If contact is made, *immediately* wash your skin and flush your eyes with clean, cold water. Then, *get medical attention.*

Clear Casting and Encapsulating

Fig. 21-1 Clear casting and encapsulating equipment and materials.

Fig. 21-2 Applying mold release agent.

Fig. 21-3 Objects to be encapsulated.

8. Learn the safe operation of the electric buffing machine.

EQUIPMENT AND MATERIALS

The equipment and materials needed for clear casting and encapsulating are:

1. Safety glasses.
2. Disposable gloves.
3. Mold.
4. Mold release (paste wax).
5. Resin and catalyst.
6. Coloring agent (optional).
7. Objects to be clear cast and encapsulated.
8. Several large, unwaxed paper containers.
9. Mixing sticks.
10. Acetone and container.
11. 100 through 800 grit wet/dry sandpaper.
12. Sanding block.
13. Buffing wheel.
14. Buffing machine.
15. Buffing compound.
16. Small sheet of glass.
17. Artist's brush.
18. Tweezers.
19. Pen.
20. Clean cloth or paper towels.
21. Weight scales

CLEAR CASTING AND ENCAPSULATING LAB PROCEDURE

1. Obtain all the items listed under "Equipment and Materials" needed to make a casting. See Fig. 21-1.

2. Select the mold for this assignment. Molds are made from aluminum, polyethylene, RTV silicone, polyurethane, latex, and ceramic. The mold used in this assignment is made of aluminum.

3. Apply two to three coats of mold release. The release agent is placed on the inside mold surface. See Fig.

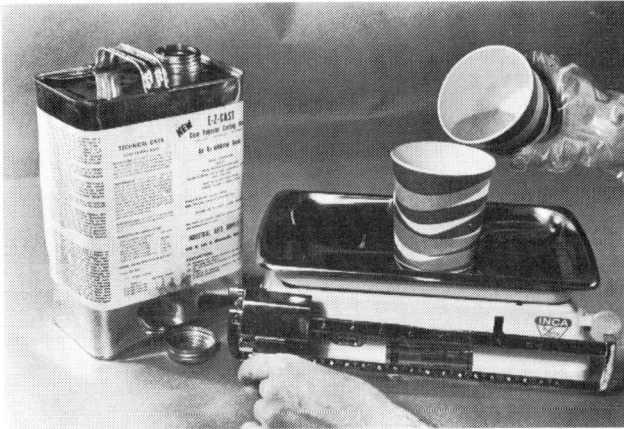

Fig. 21-4 Weighing out the resin.

Fig. 21-5 Adding catalyst drops.

21-2. Be sure to buff each coating of release agent. Apply the agent according to your instructor's directions. Be sure to remove all particles of release agent from fine detailed mold areas. Polyethylene and glazed ceramic molds often require no release agent.

4. Obtain all the objects to be encapsulated. See Fig. 21-3.

5. Clean each object to be encapsulated. Make sure each object is dry before encapsulating it. This is especially true for insects and plants. Butterfly bodies are often removed and replaced with paper bodies. This prevents the casting heat (exotherm) from burning them.

6. Weigh enough resin to cast an ⅛" layer in the mold. See Fig. 21-4. Ask your instructor how much resin to weigh out.

7. Add the proper number of catalyst drops to the resin. See Fig. 21-5. Ask your instructor how many drops to use.

8. Mix the catalyst and resin well. Try not to trap air in the resin as you mix. Follow your instructor's mixing directions. Be sure to wear disposable gloves when handling the resin and catalyst. Work in a well ventilated area.

9. Pour the catalyzed resin slowly into the mold. See Fig. 21-6. Let the resin flow slowly over the mold bottom. This prevents air bubbles from being trapped on the surface of the casting.

10. Lightly tap or vibrate the cast layer of resin before it gels. This helps air bubbles float to the surface.

11. Break any air bubbles that rise to the resin surface. See Fig. 21-7. Use a nail or pin.

Clear Casting and Encapsulating

12. Let the resin cure to a gel (sticky) stage. Make sure it is sticky enough to support the object to be encapsulated.

13. Catalyze enough resin for the next ⅛″ casting layer. Use the procedure demonstrated in Steps 6, 7, and 8.

14. Brush a layer of catalyzed resin over the surface of the object. See Fig. 21-8. This prevents air bubbles from being trapped on and around the object.

15. Lay the resin-coated object on the first cast layer.

16. Pour the second ⅛″ layer of catalyzed resin slowly over the first layer and the object. See Fig. 21-9.

17. Let the second layer cure to a gel stage.

18. Catalyze enough resin for the next ⅛″ casting layer. Use the procedure demonstrated in Steps 6, 7, and 8.

19. Continue catalyzing, pouring and gelling ⅛″ layers of resin until the mold is filled.

20. Place a clean sheet of glass over the mold after the last resin layer is cast. See Fig. 21-10. Make sure the glass touches the wet resin surface. The glass will make a shiny surface on the casting.

21. Let the casting cure. The casting will make heat (exotherm) as it cures. As the casting starts to cure, a few surface stress bumps will form on the surface. The stress marks will disappear when the casting totally cures.

22. Take the glass off the mold.

23. Touch the casting surface. If the surface is not sticky, it is cured.

Fig. 21-6 Pouring catalyzed resin into mold.

Section VI: Polyester Molding

Fig. 21-7 Breaking air bubbles.

Fig. 21-8 Brushing on catalyzed resin layer.

Fig. 21-9 Pouring second layer of catalyzed resin.

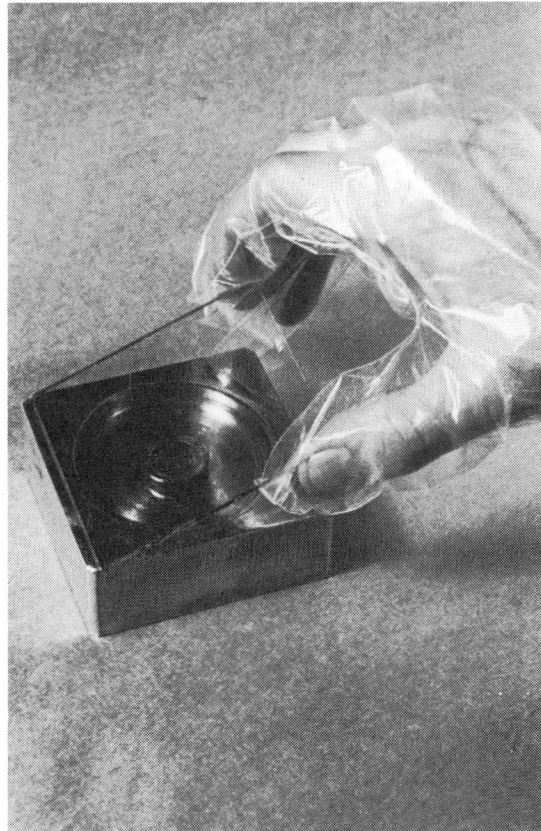

Fig. 21-10 Placing glass sheet over mold.

Clear Casting and Encapsulating

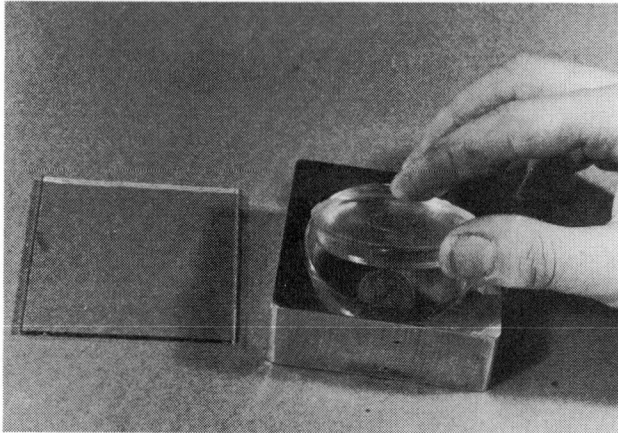

Fig. 21-11   Removing casting from mold.

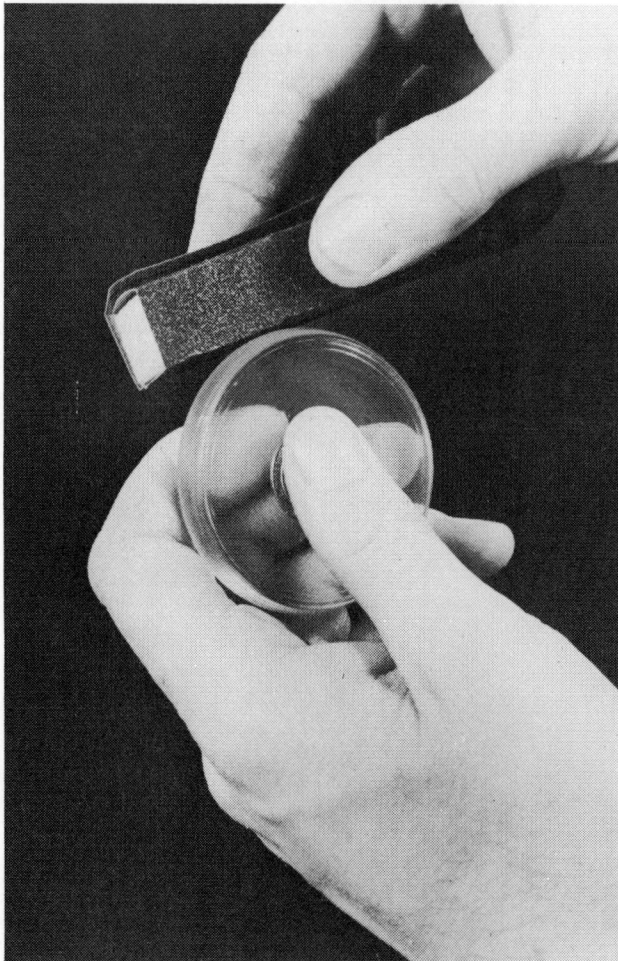

Fig. 21-12   Removing sharp casting edges.

24. Remove the casting from the mold. See Fig. 21-11.

25. Remove the sharp casting edges by sanding the edges. See Fig. 21-12. Sand with a coarse grit, wet or dry paper. Then finer grit papers are used until a 600 or 800 grit finish is made. Be sure to use a sanding block with the sandpaper. This helps keep the edges from being sanded round.

26. Buff the sanded edges by hand or with a buffing wheel. See Fig. 21-13. Be careful not to drop the hard casting. Also, be careful not to burn the casting with the buffing wheel. Use a buffing compound recommended by your instructor.

CONCLUSION

Many encapsulated objects and castings are shown in Fig. 21-14. Use other molds and objects and the procedure described above to make these items.

The process just described used polyester casting resin. This resin is very popular because it will stay clear for years. If polyester laminating resin is used by mistake, the casting may turn pink. Resins other than polyester are now being used. A few of these include acrylic, epoxy, RTV silicone, phenolic, polyurethane, vinyl, and water extended polyester.

Often, cast products are not seen by the consumer. An example is the windings of an electric motor. They have plastic cast around them under vacuum (potting).

Casting and encapsulating have several advantages. The molds are economical, low cost equipment is used, and only a small number of casting steps are needed.

The cured casting will have a

concave surface next to the glass mold cover. Also, the resin shrinks when curing. Sometimes a casting will crack, froth, discolor, or get extremely hard. This means that the catalyst to resin ratio was wrong. If too much catalyst is used for each layer, a great amount of heat (exotherm) is developed. The heat often causes the above defects.

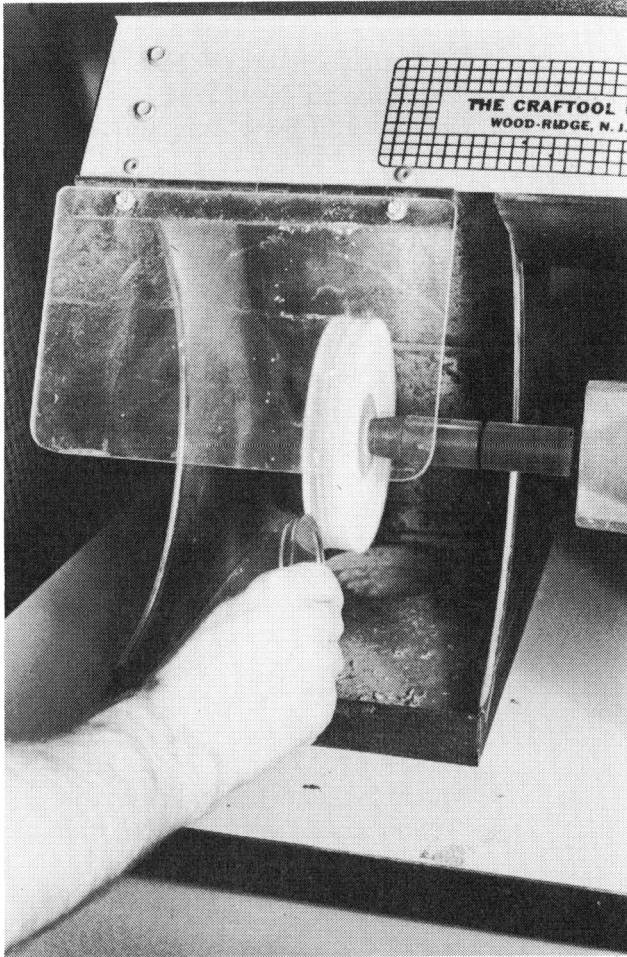

Fig. 21-13 Buffing the sanded edges.

Fig. 21-14 Examples of encapsulated objects and castings.

REVIEW

1. List four products made by clear casting and encapsulating.

2. Name three materials used for making casting molds.

3. Describe the resin-to-catalyst ratio used for your casting.

4. Describe each step in the clear casting and encapsulating processes.

5. Name the material from which your mold was made.

6. Explain how polyester resin cures.

7. Name the mold release agent you used on your mold.

8. Name two resins other than polyester used for plastic casting.

9. Explain why "potting" is not the same as encapsulating.

10. List three casting defects caused by adding too much catalyst to the resin.

11. Explain why wet samples must be dehydrated (dried) before encapsulating them.

12. Describe how to remove and avoid air bubbles in a curing casting.

Clear Casting and Encapsulating

13. Describe how to trim, sand, and polish a cured casting.

SELECTED BIBLIOGRAPHY

Agranoff, Joan, ed. *Modern Plastics Encyclopedia.* New York: McGraw-Hill Book Company, 1976-77.

Baird, Ronald J. *Industrial Plastics.* South Holland, Illinois: The Goodheart-Willcox Company, Inc., 1971.

Casting and Encapsulating—Teacher's Manual. Indianapolis: Howard W. Sams and Company, Inc., 1974.

Holt, Elmer. *Casting in Clear Plastic.* Temple City, California: 1970.

Milby, Robert V. *Plastics Technology.* New York: McGraw-Hill Book Company, 1973.

Patton, William J. *Plastics Technology: Theory, Design, and Manufacture.* Reston, Virginia: Reston Publishing Company, Inc., 1976.

Richardson, Terry A. *Modern Industrial Plastics.* Indianapolis: Howard W. Sams & Company, Inc., 1974.

Rosato, Dominick V., ed. *Plastics Industry Safety Handbook.* Boston: Cahners Books, 1973.

EQUIPMENT AND MATERIAL SUPPLIERS

1. Ain Plastics, Inc., 65 Fourth Avenue, New York, New York 10003.

2. Brodhead-Garrett, 4560 East 71st Street, Cleveland, Ohio 44105.

3. Cope Plastics, Inc., 4441 Industrial Drive, Godfrey, Illinois 62035.

4. Delvies Plastics, Inc., 2320 South West Temple, P.O. Box 1415, Salt Lake City, Utah 84110.

5. Industrial Arts Supply Company, 5724 West 36th Street, Minneapolis, Minnesota 55416.

6. McKilligan Industrial Supply Corporation, 494 Chenango Street, Binghamton, New York 13901.

The following products may be purchased from the suppliers designated by the code numbers used above.

Casting resin and colorants can be purchased from any of the above listed suppliers. Molds are available from suppliers 2, 3, 4, 5 and 6.

Clear Casting and Encapsulating

Assignment 22
Water Extended Polyester

OBJECTIVES

To prepare a mold.

To water extend polyester.

To make a W.E.P. casting.

INTRODUCTION

Water extended polyester casting is a water and resin mixture cast into a mold. After curing, the casting looks like plaster. Water extended polyester is a composite (a mixture of materials) often used for furniture parts, art objects, and small statues.

The casting materials are generally mixed one part resin to one part water. If more water is mixed, the strength of the casting will decrease. Water extended polyester is used as a casting material. After the material cures, it may be machined, nailed, sawed, sanded, painted, and glued. Fillers (glass fibers) are often added to the resin and water mix to make a stiff, high density casting.

Water is the basic bulking material in W.E.P. It also aids in getting rid of the heat (exotherm) developed in the casting. The water stays in the casting almost indefinitely.

SAFETY

The following precaution should be taken when casting with water extended polyester (W.E.P.):

1. Work in a well-ventilated area.

2. Wear safety glasses and disposable gloves.

3. Keep a general purpose fire extinguisher in the work area.

4. Keep resin away from heat and flame.

5. Be sure to measure the catalyst correctly before you add it to the resin.

6. Learn the safe operation of the electric mixing equipment.

7. Protect your eyes and skin from contact with the resin. If contact is made, *immediately* wash your skin and flush your eyes with clean, cold water. Then, *get medical attention.*

EQUIPMENT AND MATERIALS

The equipment and materials needed for casting with water extended polyester (W.E.P.) are:

1. Safety glasses.
2. Disposable gloves.
3. Mixer (electric drill or press).
4. Mixing blade (high shear and minimum air entrapment design).
5. Mold.
6. Mold release agent (paste wax).
7. Resin (W.E.P.) and catalyst.
8. Coloring agent (optional).
9. Fillers (fiber glass, old W.E.P. casting particles) (optional).
10. 3 large, unwaxed paper containers.
11. Mixing sticks.
12. Jar.
13. Water.
14. Weight scale.
15. Clean cloth or paper towels.
16. Acetone.

WATER EXTENDED POLYESTER CASTING LAB PROCEDURE

1. Obtain all the items listed under "Equipment and Materials" needed

Fig. 22-1 Water extended polyester casting equipment and materials.

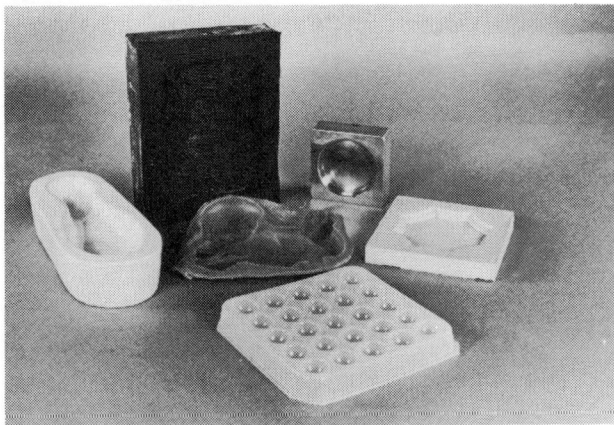

Fig. 22-2 Molds for water extended polyester casting.

Fig. 22-3 Applying the release agent.

to make a W.E.P. casting. See Fig. 22-1.

2. Select the mold for this assignment. Molds are made from aluminum, polyethylene, RTV silicone, polyurethane, latex, and ceramic. See Fig. 22-2. The mold used in this assignment is latex in a wood and plaster retainer. See Assignment 12 to make latex molds. Molds give W.E.P. casting its final shape. RTV silicone, polyurethane, and latex molds are used when fine details must be cast.

3. Apply two to three coats of mold release. Place the release agent on the inside mold surface. See Fig. 22-3. **Be sure to buff each coating of release agent.** Apply the agent according to your instructor's directions. **Be sure to remove all particles of release agent from fine detailed mold areas.** Polyethylene, polished aluminum, and glazed ceramic molds often require no release agent.

4. Calculate the volume of the mold and add 5% for waste. Your instructor will help you determine the mold volume.

5. Pour water equal to *half* the volume of the mold into a cup.

6. Pour polyester equal to *half* the volume of the mold into a large container. See Fig. 22-4. Make sure the container is large enough to hold the resin *and* the water. One part water is generally mixed with one part resin by volume. This is the mixing ratio for extending polyester.

7. Place a mixing blade in an electric drill.

8. Place the mixing blade deep in the resin.

9. Turn on the drill. See Fig. 22-5. Run the mixing blade at 1500 to

Section VI: Polyester Molding

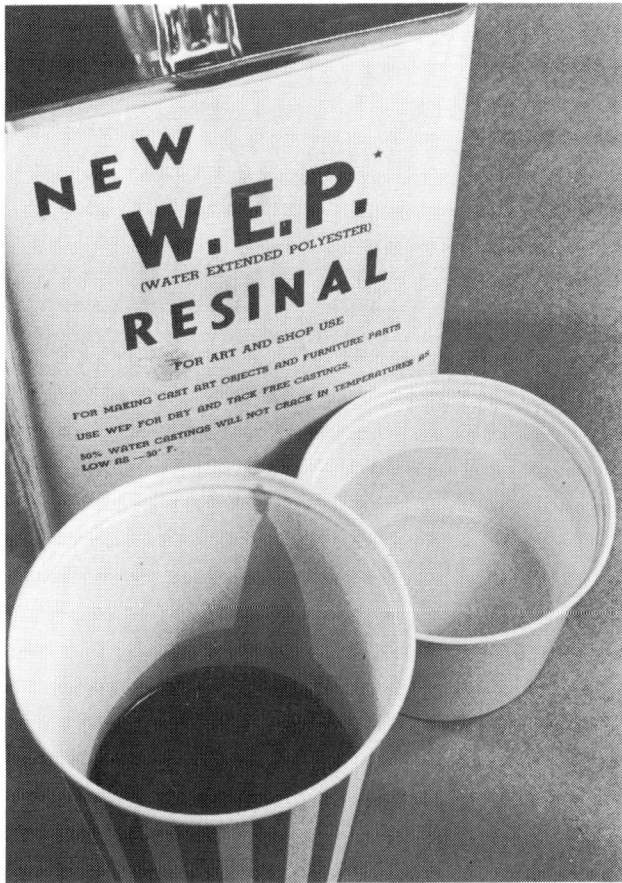

Fig. 22-4 Pouring the polyester.

2000 RPM. A vortex (funnel) should form in the resin. Make sure the mixing blade is not blocked. Hold the container tight.

10. Slowly let the water drip down the mixing blade shaft. See Fig. 22-6. The slower the water is added to the resin, the better the resin and water are mixed. Never add resin to the water. Let the mixed W.E.P. stand a few minutes. This lets air escape from the mixture. Clean the mixing blade with acetone immediately.

11. Weigh the correct amount of catalyst (MEK) for the weight of W.E.P. to be cast. See Fig. 22-7. Your instructor will tell you how much catalyst to weigh out. The catalyst used is less than 1% of the W.E.P. weight.

12. Pour the catalyst into the W.E.P. container. See Fig. 22-8.

13. Mix the catalyst and W.E.P. well. Follow your instructor's mixing

Fig. 22-5 Mixing with an electric drill.

Fig. 22-6 Dripping water down blade shaft.

Water Extended Polyester (WEP)

Fig. 22-7   Weighing out the catalyst.

Fig. 22-8   Pouring catalyst into W.E.P. container.

instructions. Be sure to completely mix the catalyst and the W.E.P. Wear disposable gloves when handling catalyst and resin. Work in a well ventilated area.

14. Pour the catalyzed W.E.P. slowly into the mold. See Fig. 22-9. Let the W.E.P. flow slowly over the mold surface. This prevents air bubbles from being trapped on the surface of the casting.

15. Let the casting cure in the mold at room temperature. The casting gels within 10 minutes. The casting will make heat (exotherm) as it cures.

16. Remove the cured casting from the mold. Castings can often be removed 45 minutes after pouring. Be careful not to break any thin sections of the casting.

17. Look at the finished W.E.P. casting in Fig. 22-10. The surface is smooth and the details clear. Some castings have a pock-marked surface. This is caused by poor water-to-resin mixing. It is also caused by silicone release agents.

CONCLUSION

Many W.E.P. castings are shown in Fig. 22-11. Use other molds and the

Fig. 22-9   Pouring catalyzed W.E.P. into mold.

Section VI: Polyester Molding

procedure just described to make these castings.

Water extended polyester resin is very popular. This results from its being one of the cheapest casting resins. The resin is extended by 50% with water. This means the original volume has been doubled. This lowers the cost per casting.

An uncatalyzed water and resin emulsion can be stored for two days. It must then be catalyzed and used. After the W.E.P. is cast and cured, it will undergo a slight amount of shrinkage. This is caused by water loss.

To produce high quality castings, the water must be added and mixed in small droplets. High shear mixers are used. If this type of mixing is not done, the casting will be composed of large particle emulsions. This greatly reduces the casting strength. Remember that an undermixed solution or emulsion is very runny (low viscosity).

REVIEW

1. Name two products made by W.E.P. casting.

2. Describe the water-to-resin and the W.E.P.-to-catalyst ratios used for your casting.

3. Name three materials used for making W.E.P. molds.

4. Explain the reason for using water in a W.E.P. emulsion.

5. Demonstrate how to emulsify resin and water.

6. Explain why glass fibers are mixed into uncatalyzed but emulsified resin.

7. Explain what an emulsified composite is.

Fig. 22-10   Finished W.E.P. casting.

Fig. 22-11   Examples of W.E.P. castings.

Water Extended Polyester (WEP)

8. Explain why W.E.P. resin is cheaper than other casting resins.

9. Demonstrate an exothermic reaction.

10. Name two causes of pock-marked castings.

SELECTED BIBLIOGRAPHY

Casting and Encapsulating—Teacher's Manual. Indianapolis: Howard W. Sams & Company, Inc., 1974.

Richardson, Terry A. *Modern Industrial Plastics.* Indianapolis: Howard W. Sams & Company, Inc., 1974.

Rosato, Dominick V., ed. *Plastics Industry Safety Handbook.* Boston: Cahners Books, 1973.

EQUIPMENT AND MATERIAL SUPPLIERS

1. Delvies Plastics, Inc., 2320 South West Temple, P.O. Box 1415, Salt Lake City, Utah 84110.

2. Industrial Arts Supply Company, 5724 West 36th Street, Minneapolis, Minnesota 55416.

W.E.P., colorants, and molds can be purchased from either of the suppliers listed above.

Section VII
Foam
Molding

Assignment 23
Polystyrene Expandable Beads

OBJECTIVES

To prepare expandable bead foaming equipment.

To pre-expand beads.

To make a polystyrene expandable bead foamed product.

INTRODUCTION

Polystyrene expandable bead foaming is a high speed industrial process. It is used to make plastic products. These products are buoyant, shock absorbing, insulating, and lightweight. A few of these products include ice chests and buckets, hot and cold cups, packaging containers, flotation devices, and toys. An industrial cup molding machine is shown in Fig. 23-1.

The polystyrene expandable bead foaming process works as follows. Small, gas filled (generally penthane), polystyrene beads are heated in a container. The heat softens the bead wall. It also causes the gas inside the bead to expand. Because of this, the whole bead expands. This operation is called "pre-expansion." Often the beads are pre-expanded to 50% of their part volume. The longer the beads are left in the pre-expander container, the bigger they get.

A product made with large beads will have a low density. The same product made with many more small beads will have a high density. The future use of a product determines whether large or small beads are used. The high density product will weigh more than the low density product.

Fig. 23-1   Industrial cup molding machine. (Courtesy Uniloy-Springfield Division/Hoover Ball & Bearing Company)

After pre-expansion, the beads are placed in a mold. The mold, filled with the pre-expanded beads, is closed and heated with hot water or steam. The heating causes the pre-expanded beads to heat again. This causes the bead walls to soften and the gas inside to expand. Because the beads are trapped inside the mold, the softened walls of the bead touch other bead walls. This makes them stick together. All the beads fuse to form one product. The hot mold and product are cooled. Then, the cooled product is removed from the mold.

SAFETY

The following precautions should be taken when polystyrene expandable bead foaming:

1. Work in a well-ventilated area. The bead storage container will give off volatile fumes when opened. These fumes are formed during storage.

2. Work on a heat-resistant surface.

3. Wear safety glasses and heat-resistant gloves.

4. Keep a general purpose fire extinguisher in the work area.

5. Do not touch the hot molds, water container, or equipment with your bare hands.

6. Do not burn yourself with the hot air gun.

7. Do not burn yourself on the hot water or steam made during the foaming process.

8. Be sure to use polystyrene expandable bead foam plastic according to the manufacturer's instructions.

9. Learn the safe operation of the hot plate.

Polystyrene Expandable Beads

10. Learn the safe operation of the bead pre-expander.

EQUIPMENT AND MATERIALS

The equipment and materials needed for polystyrene expandable bead foaming are:

1. Safety glasses.
2. Heat-resistant gloves.
3. Heat-resistant surface.
4. Molds.
5. Mold release agent.
6. Unexpanded polystyrene beads.
7. Water boiling container.
8. 16-oz. cup.
9. Sieve.
10. Funnel.
11. Trimming board.
12. Utility knife.
13. Pliers.
14. Screwdriver.
15. Clean cloth or paper towels.
16. Hot plate.

POLYSTYRENE EXPANDABLE BEAD FOAMING LAB PROCEDURE

A polystyrene expandable bead foaming setup for a school shop is shown in Fig. 23-2. Your shop may have a similar one.

1. Obtain all the items listed under "Equipment and Materials" needed for polystyrene expandable foaming. See Fig. 23-2.

2. Heat a container of water (¾ full) to a boil.

3. Pour 2 or 3 ounces of unexpanded beads into the boiling water. See Fig. 23-3. At first, the beads will sink. As the gas expands inside the beads, they will float.

4. Remove the beads with a sieve when they reach the size of

Fig. 23-2 Polysytrene expandable bead foaming equipment and materials.

Fig. 23-3 Pouring unexpanded beads into boiling water.

Fig. 23-4 Removing beads with sieve.

Fig. 23-5 Beads placed for drying.

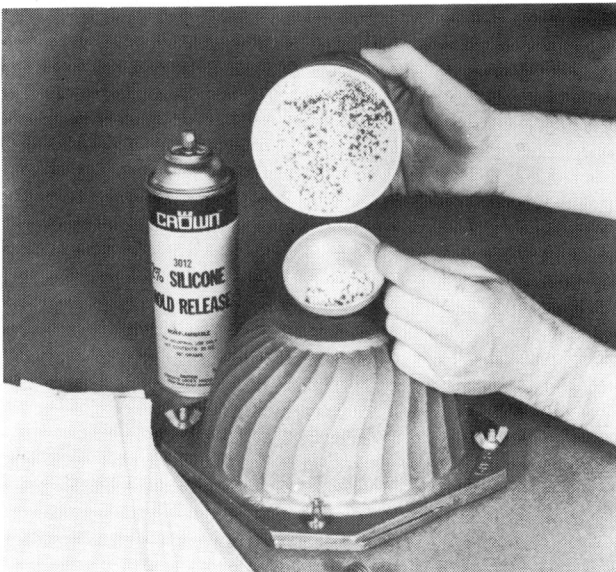

Fig. 23-6 Filling mold with pre-expanded beads.

beebees. See Fig. 23-4. The pre-expanded beads should now be about 50% of their final size. The bead pre-expansion controls the final density of the product.

5. Place the wet beads on a surface to dry. See Fig. 23-5. The wet beads may be placed in the mold or stored in a plastic bag. **Do not pre-expand more beads than you can use in two days.**

6. Wipe the inside mold surfaces clean.

7. Place the release agent on the inside mold surfaces.

8. Clamp the mold halves together. Use bolts and wing nuts.

9. Completely fill the mold with the pre-expanded beads. See Fig. 23-6.

10. Replace the mold filler hole plug. Polystyrene expandable bead foaming molds are generally made of aluminum.

11. Place the filled mold in a container of boiling water. See Fig. 23-7. Ask your instructor how long to heat the mold. Be careful not to get a steam or hot water burn. The time needed to expand the beads depends on the final thickness of the product. The thicker the product, the longer it will take to foam it. Thick parts must stay in the boiling water longer than thin ones.

12. Cool the mold in cold water. See Fig. 23-8. **Make sure the mold and product are completely cooled.** If the mold is opened while it or the product is warm, the product will continue to expand.

13. Remove the mold clamp nuts and bolts.

14. Pry the mold halves apart by hand.

Polystyrene Expandable Beads

Fig. 23-7   Placing filled mold in boiling water.

Fig. 23-8   Cooling mold in cold water.

Fig. 23-9   Removing product from mold.

Fig. 23-10   Trimming flash.

Fig. 23-11   Finished Bowl.

Section VII: Foam Molding

15. Remove the product from the mold. See Fig. 23-9. **Do not touch the inside of the mold with a tool.** This will scratch the mold surfaces.

16. Trim the flash from the product with a sharp utility knife. See Fig. 23-10. This step is often not needed. Make sure the knife is sharp. Be careful not to cut yourself.

17. Look at the finished product. See Fig. 23-11. The surface should be smooth and free of flash. It can be used as a planter. With a lid, it can also be used as an ice bucket or bowl.

CONCLUSION

Fig. 23-12 shows many products that were "foamed" with polystyrene expandable beads. Different molds were used, but the procedure just described was followed.

The process described above is slow. To increase bead expansion production, industry continuously pre-expands beads with automated machines. One type is a radiant heat pre-expander. With this machine, the unexpanded beads feed from a hopper onto a moving belt. The belt passes the beads under or through a heat source.

As the beads pass through the heat source, they are pre-expanded. The amount of bead pre-expansion is determined by the speed of the belt. It is also determined by the amount of heat used. Beads pass from the belt into storage containers.

The final bead foaming operation is a continuous process. Pre-expanded beads are first removed from storage containers. They are blown into closed molds in large automated molding machines. Steam is passed through the molds. This causes the beads to expand and fuse into finished products. The molds and parts are then allowed to

Fig. 23-12 Examples of polystyrene expandable bead foamed products.

Polystyrene Expandable Beads

cool. The parts are removed from the molds, trimmed, and packaged. You have probably seen many products that were made this way.

REVIEW

1. List three products made by polystyrene expandable bead foaming.

2. Describe each step in polystyrene expandable bead foaming.

3. Describe how to continuously pre-expand polystyrene beads.

4. Explain why polystyrene beads are pre-expanded.

5. Describe how industry continuously foam molds polystyrene bead products.

6. Name the material from which the lab activity product mold was made.

7. State the percent of final volume into which each bead was pre-expanded.

8. Describe how to calculate the amount of time the charged mold will be in the boiling water.

9. Explain why the mold and part must be thoroughly cooled before opening the mold.

SELECTED BIBLIOGRAPHY

Agranoff, Joan, ed. *Modern Plastics Encyclopedia.* New York: McGraw-Hill Book Company, 1976-77.

Baird, Ronald J. *Industrial Plastics.* South Holland, Illinois: The Goodheart-Willcox Company, Inc., 1971.

Foam Molding-Teacher's Manual.
Indianapolis: Howard W. Sams &
Company, Inc., 1974.

Milby, Robert V. *Plastics Technology.*
New York: McGraw-Hill Book
Company, 1973.

Patton, William J. *Plastics
Technology: Theory, Design, and
Manufacture.* Reston, Virginia;
Reston Publishing Company, Inc.,
1976.

Richardson, Terry A. *Modern
Industrial Plastics.* Indianapolis:
Howard W. Sams & Company, Inc.,
1974.

Rosato, Dominick V., ed. *Plastics
Industry Safety Handbook.* Boston:
Cahners Books, 1973.

EQUIPMENT AND MATERIAL SUPPLIERS

1. Delvie's Plastics, Inc., 2320 South
West Temple, P.O. Box 1415, Salt
Lake City, Utah 84110.

2. Industrial Arts Supply Company,
5724 W. 36th St., Minneapolis,
Minnesota 55408.

3. McKilligan Industrial Supply
Corporation, 494 Chenango Street,
Binghamton, New York 13901.

The plastic and molds used in the
polystyrene expandable bead
foaming process can be purchased
from any of the suppliers listed
above. The pre-expander and cooker
are available from suppliers 2 and
3.

Assignment 24
Polyurethane Foam

OBJECTIVES

To prepare a mold.

To prepare polyurethane foaming equipment and plastic.

To make a polyurethane foamed product.

INTRODUCTION

Polyurethane chemical foaming produces rigid (closed-cell) and flexible (open-cell) plastic products. Among the many examples of these products are furniture stuffing material, bedding material, water skiing belts and jackets, boat flotation chamber interiors, foamed-in-place insulation, imitation wood hand-hewn interior ceiling beams, and some furniture parts.

Polyurethane foaming is done as follows. A mold is opened and a mixed amount of foam is placed into one mold half. The other mold is closed over the charged half. Both halves are tightly clamped. The mixed chemicals react and make heat. The heat expands a gas foaming agent (generally freon) found in one part of the two part system. As the gas volume increases, it expands the plastic.

The expanded plastic is filled with bubbles. The heat produced during foaming causes the bubbly foam to solidify. Foaming and curing are then complete. The part is removed from the mold.

A small, industrial-type polyurethane foam operation is shown in Fig. 24-1. It is a ball mold. It will be filled with foam from the operator's gun.

Fig. 24-1 Industrial polyurethane foam operation. (Courtesy Gusmer Corporation)

Section VII: Foam Molding

SAFETY

The following precautions should be taken when working with polyurethane foam:

1. Work in a well-ventilated area. This allows volatile fumes produced during the foaming process to escape.

2. Avoid breathing the toxic polyurethane component vapors.

3. Wear a respirator when machining urethane foam. This reduces your chance of breathing urethane particles.

4. Work on a heat-resistant surface.

5. Wear heat-resistant gloves when handling hot molds.

6. Wear disposable gloves when handling and mixing polyurethane foam components.

7. Wear safety glasses.

8. Keep a general purpose fire extinguisher in the work area.

9. Wash thoroughly with soap and water if *any* urethane foam components contact your skin. (Especially component **A.**)

10. Learn the safe operation of the mold preheating oven.

11. Learn the safe operation of the electric drill and the mixing attachments.

12. Be sure to use all polyurethane components according to the manufacturer's instructions.

EQUIPMENT AND MATERIALS

The equipment and materials needed for polyurethane foaming are:

Polyurethane Foam

1. Safety glasses.
2. Heat-resistant gloves.
3. Disposable gloves.
4. Heat-resistant surface.
5. Vented industrial type or laboratory type preheating oven with heat circulating fan.
6. Mold.
7. Mold clamping device (fixture or clamps).
8. Mold release agent.
9. Polyurethane foam components.
10. Mold board.
11. Plastic release sheet (polyethylene).
12. Container (1 or 2 lb. coffee can).
13. 2 large, unwaxed paper cups.
14. ⅜" variable speed electric drill with mixing attachment.
15. Mixing stick.
16. Weight scales.
17. Scissors.
18. Clean cloth or paper towels.
19. Newspapers.

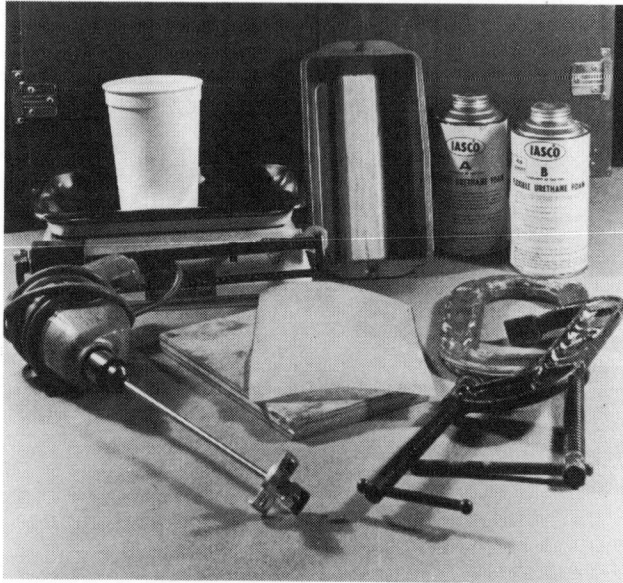
Fig. 24-2 School laboratory polyurethane foaming setup.

Fig. 24-3 Polyurethane foaming equipment and materials.

POLYURETHANE FOAMING LAB PROCEDURE

A school laboratory polyurethane foaming setup is shown in Fig. 24-2. This may be similar to the equipment in your school shop. Take a moment to identify each of the items in the picture. The polyurethane foaming procedure is as follows:

1. Obtain all the items listed under "Equipment and Materials" needed to polyurethane foam a product. See Fig. 24-3.

2. Obtain the two polyurethane foam components.

3. Select the mold for this assignment. Molds are made from aluminum, sheet metal, RTV silicone, and heat-resistant plaster. See Fig. 24-4. The cartop carrier mold used in this assignment is made of sheet steel.

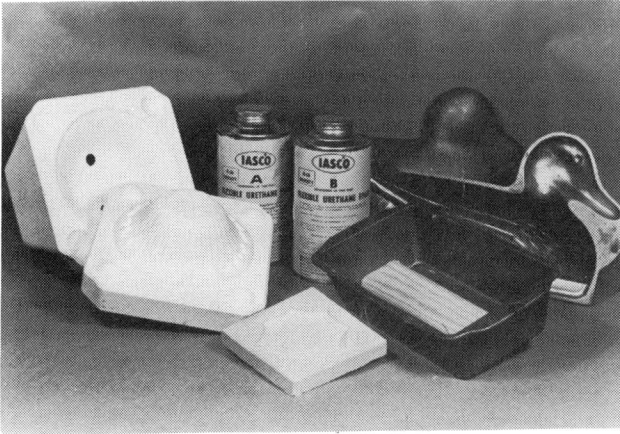

Fig. 24-4 Typical polyurethane molds and foaming materials.

Fig. 24-5 Applying the release agent.

Fig. 24-6 Preheating metal mold.

4. Calculate the volume of the mold. Fill the mold with sand or water. Pour the mold sand or water into a volume calibrated (marked) container. This tells you the size of the space (volume) taken up by the mold. Be sure to wipe out the mold before going to the next step.

5. Apply two to three coats of mold release. A high quality furniture paste wax can be used. Perma-Mold Release No. 2-27 from Cope Plastics, Inc. can also be used. The release agent is placed on the inside surfaces and on the plastic mold top sheet. See Fig. 24-5. Be sure to buff each coating of release agent. Apply the release agent according to your instructor's directions. Different molds use different release agents. The type used depends on the mold material. It also depends on the finish (paint, etc.) to be placed on the foamed part.

6. Preheat the metal molds from 90 to 120° F. See Fig. 24-6. Some molds do not need preheating. Follow your instructor's directions for mold preheating.

7. Weigh out the amount of foam (parts A and B) needed to fill the volume calculated in Step 4. See Fig. 24-7. Ask your instructor to explain the procedure for making

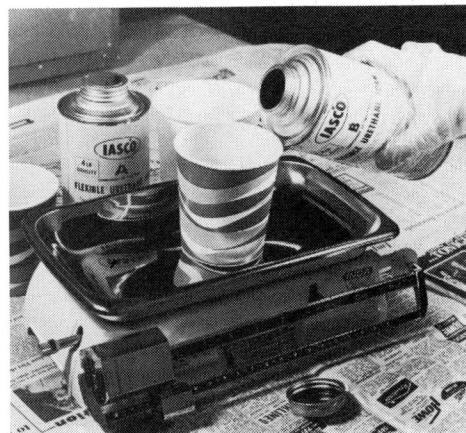

Fig. 24-7 Weighing out the foam.

Polyurethane Foam

the calculations. Rigid foam expands to about 20 times its original volume. Flexible foam expands to about 15 times its original volume. Be sure to use equal weights or volumes of part A and part B.

8. Keep the equal parts of A and B in separate, unwaxed containers.

9. Pour parts A and B together into a separate, large, unwaxed container.

10. Mix parts A and B together for about 20 to 30 seconds. Mix by hand or with an electric drill and a mixing attachment. See Fig. 24-8. Run the drill at 300-400 RPM. Follow your instructor's foam mixing instructions. Be sure to completely mix parts A and B. Stop mixing before the foam starts to expand. Be sure to immediately clean the mixing materials with acetone.

11. Pour the mixed foam into the mold. See Fig. 24-9. Make sure all the foam is in the mold before it expands. Fig. 24-10 shows the foam expanding in the mold.

12. Lay the plastic mold top release sheet and wood on the open end of the mold.

13. Clamp the plastic sheet and wood top securely to the mold. See Fig. 24-11. The clamps should be adjusted and ready before the foam is mixed.

14. Unclamp the mold top after the foam has set up. Some foams set up in about fifteen minutes. Ask your instructor how much time to allow. Carefully hand flex the product from the mold. See Fig. 24-12.

15. Place the foamed product in the oven to cure. See Fig. 24-13. Ask your instructor what heating temperature and time to use. This step will not be needed for certain foams.

Fig. 24-8   Mixing with electric drill.

Fig. 24-9   Pouring mixed foam into mold.

Fig. 24-10 Foam expanding in mold.

Fig. 24-11 Release sheet and wood clamped to mold.

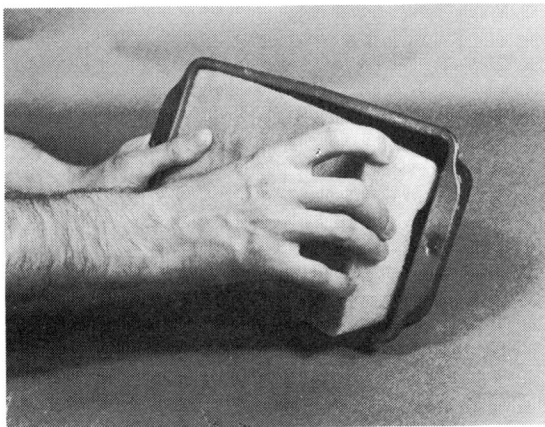
Fig. 24-12 Removing product from mold.

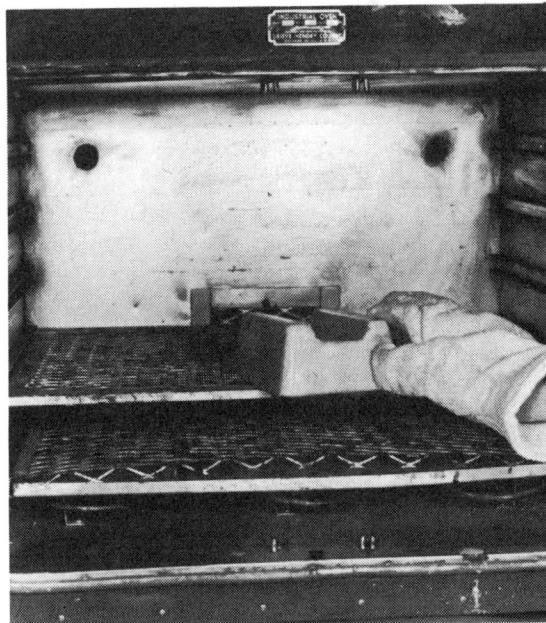
Fig. 24-13 Curing the foamed product.

Polyurethane Foam

16. Cut the flash from the product with scissors. Trim rigid flash with a utility knife. If you sand rigid foam, wear a respirator. **Do not cut the foam until all the reaction heat is gone.**

CONCLUSION

Many products that were polyurethane foamed are shown in Fig. 24-14. To make other products, use a different mold and follow the procedure just described.

The foaming process described above is slow, but adequate for laboratory work. Foam product production can be greatly increased. Industry often uses special equipment to continuously produce foamed products. One piece of equipment resembles a large spray gun. Foam parts are automatically mixed in the spray gun. They are then shot into open molds. This is how some foamed furniture parts or imitation wood, hand-hewn ceiling beams are made.

Similar equipment is used to coat certain building materials. The equipment continuously sprays a foam coating on the backs of the materials. It is also used to spray foam insulation onto industrial and residential building walls.

REVIEW

1. List four products made by the polyurethane foam process.

2. Describe each step in the polyurethane foam process.

3. Describe one industrial method of continuously making polyurethane foamed products.

4. Name the material used to construct your mold.

Fig. 24-14 Examples of polyurethane foamed products.

Section VII: Foam Molding

5. Explain the principle of polyurethane chemical foam molding.

6. Explain what the chemical foam molding blowing agent does.

7. Name two uses for open-celled and closed-celled foamed products.

8. Explain the differences between chemical and physical plastic foaming.

SELECTED BIBLIOGRAPHY

Agranoff, Joan, ed. *Modern Plastics Encyclopedia.* New York: McGraw-Hill Book Company, 1976-77.

Baird, Ronald J. *Industrial Plastics.* South Holland, Illinois: The Goodheart-Willcox Company, Inc., 1971.

Foam Molding-Teacher's Manual. Indianapolis: Howard W. Sams & Company, Inc., 1974.

Kovaly, Kenneth *A. Handbook of Plastic Furniture Manufacturing.* Stamford, Connecticut: Technomic Publishing Company, Inc., 1970.

Milby, Robert V. *Plastics Technology.* New York: McGraw-Hill Book Company, 1973.

Patton, William J. *Plastics Technology: Theory, Design, and Manufacture.* Reston, Virginia; Reston Publishing Company, Inc., 1976.

Richardson, Terry A. *Modern Industrial Plastics.* Indianapolis: Howard W. Sams & Company, Inc., 1974.

Rosato, Dominick V., ed. *Plastics Industry Safety Handbook.* Boston: Cahners Books, 1973.

EQUIPMENT AND MATERIAL SUPPLIERS

1. Ain Plastics, Inc., 65 Fourth Avenue, New York, New York 10003.

2. Brodhead-Garrett, 4560 East 71st Street, Cleveland, Ohio 44105.

3. Cope Plastics, Inc., 4441 Industrial Drive, Godfrey, Illinois 62035.

4. Delvies Plastics, Inc., 2320 South West Temple, P.O. Box 1415, Salt Lake City, Utah 84110.

5. Industrial Arts Supply Company, 5724 West 36th Street, Minneapolis, Minnesota 55416.

6. McKilligan Industrial Supply Corporation, 494 Chenango Street, Binghamton, New York 13901.

Plastic for polyurethane foaming can be purchased from suppliers 1, 2, 3, 4, and 5. The mold release agent is available from suppliers 3, 4, and 5. Molds can be purchased from suppliers 3, 4, 5, and 6.

Section VIII
Coatings

Assignment 25
Cold Dip Coating
(Peelable Butyrate)

OBJECTIVES

To prepare a cutting tool edge for cold dipping.

To coat a cutting tool edge with a medium thickness, bubble-free coating of peelable butyrate.

INTRODUCTION

Coating sharpened cutting edges by using the cold dip procedure helps prevent their rusting while they are stored. The coating (peelable butyrate) seals the cutting edge from the atmosphere, preventing the cutting edge from rusting. The coating also contains an oil that lubricates the coated tool surface. The oil helps to remove the coating from the tool edge when the tool is used.

This process is called "cold dipping" because the tool is at room temperature when it is dipped. **Only the plastic is heated during the process.**

The most common plastic used as a peelable coating (in most plastics programs) is cellulose acetate butyrate (CAB). It may be bought in a variety of colors. CAB is one type of cellulosic plastic. Cellulose is made from wood pulp. Cellulosic plastics belong to the thermoplastic family of plastics. Many types of cellulose plastics, and a few vinyl-base plastics, are also used for peelable coatings.

SAFETY

The following precautions should be taken when cold dip coating:

1. Work in a well-ventilated area. Do *not* breathe the methylene chloride fumes. Also avoid breathing the CAB fumes during the coating and curing stages.

2. Work on a heat-resistant surface.

3. Wear safety glasses and heat-resistant gloves.

4. Keep a general purpose fire extinguisher in the work area.

5. Learn the safe operation of the heated butyrate melting tank.

6. Do not heat the butyrate with an open flame.

7. Do not heat the butyrate above 350° F. It may decompose above this temperature.

8. Do not cut yourself on sharp tool edges. The CAB serves as a rust-proof coating for the cutting edge. *It is not a protective sheath.*

EQUIPMENT AND MATERIALS

The equipment and materials needed for cold dip coating are:

1. Safety glasses.
2. Heat-resistant gloves.
3. Heat-resisitant surface.
4. Butyrate melting tank (industrial dip tank or deep fryer).
5. Peelable butyrate.
6. Methylene chloride.
7. Tool edge to be coated.
8. Tool support board.
9. Spray booth (optional).
10. Steel wool.
11. Clean cloth or paper towels.

COLD DIP COATING LAB PROCEDURE

1. Obtain all items listed under "Equipment and Materials" needed for cold dip coating. See Fig. 25-1.

Fig. 25-1 Cold dipping equipment and materials.

Cold Dip Coating (Peelable Butyrate)

Fig. 25-2 Setting the melt temperature adjustment switch.

Fig. 25-3 Melting material containing bubbles.

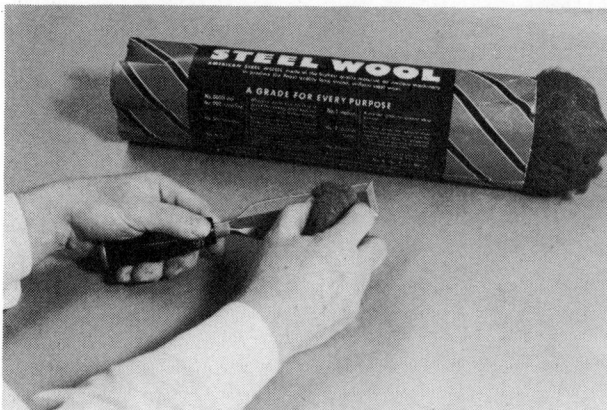

Fig. 25-4 Cleaning with steel wool.

2. Connect the peelable butyrate heating tank (deep fryer or industrial dip tank).

3. Set the melt temperature adjustment switch at approximately 350° F. See Fig. 25-2.

4. Add peelable butyrate to the melting tank if the plastic level is low. Wear heat-resistant gloves while adding the material.

6. Place the lid on the melting tank. Continue to wear heat-resistant gloves during this procedure.

7. Check the condition of the melting material every five minutes. When the liquid is free flowing and no lumps of material or bubbles remain, as in Fig. 25-3, it is ready for use. (Having melt temperature adjustment switch set at 350° F., prevents the CAB from overheating, decomposing, and igniting.)

8. Clean the sharpened cutting tool edge with steel wool to remove any dirt, scale, or rust. See Fig. 25-4. Be careful not to cut your fingers.

9. Clean the sharpened cutting tool edge with a solvent or cleaning agent such as methylene chloride. See Fig. 25-5. Be careful not to inhale the methylene chloride fumes.

10. Let the surface dry.

11. Dip the sharpened cutting tool edge into the CAB for a few seconds. See Fig. 25-6. Move the tool around in the CAB if desired. Be sure to wear heat-resistant gloves. If you want a thin coating, leave the tool edge in the CAB for a minute. If you want a thick coating, quickly dip and remove the tool edge from the CAB.

12. Remove the tool. Let the excess CAB drip into the melting pot. See

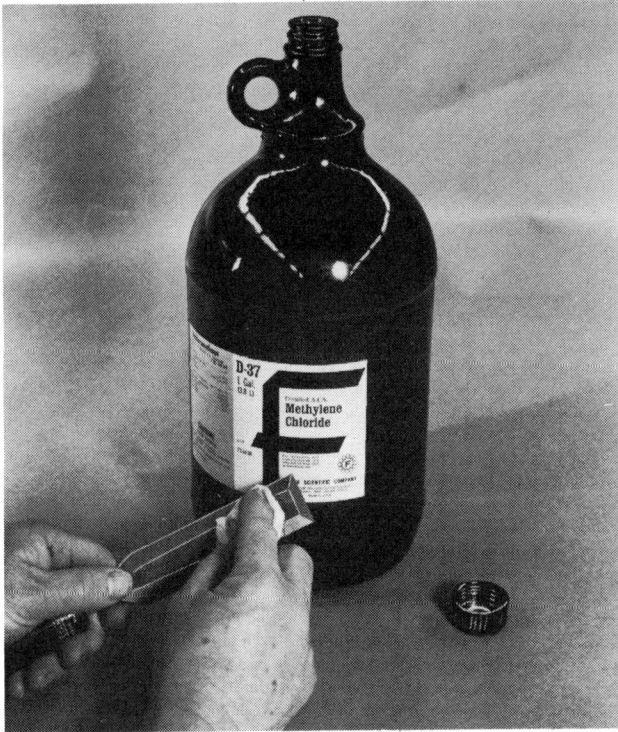

Fig. 25-5 Cleaning tool edge with methylene chloride.

Fig. 25-6 Dipping tool.

Fig. 25-7. Do not let the hot plastic contact your skin.

13. Lay the tool on a board so that the coated edge does not touch anything.

14. Let the coating cool for five minutes. The same procedure may be used to coat other types of products. See Fig. 25-8. If the part coating is extremely thin, the plastic was too hot or has decomposed. To correct this problem, add more material to the dip tank and work at a lower temperature. If the part coating is thick or bubbly, the plastic is too cold. To correct this problem, increase the dip pot temperature.

15. Remove the peelable butyrate coating before using the tool. See Fig. 25-9.

16. Wipe the oil from the coated portion of the tool.

Fig. 25-7 Allowing excess CAB to drip off tool.

Cold Dip Coating (Peelable Butyrate)

Fig. 25-8 Examples of other cold dipped coated products.

Fig. 25-9 Tool with coating removed.

Fig. 25-10 Remelting defective coating.

17. To save material, melt and reuse any defective coatings produced during the above procedure. See Fig. 25-10. Make sure the peelable butyrate is clean before placing it back into the dip tank. Remelt the scrap by using the procedure described earlier for new material.

CONCLUSION

The procedure used in the assignment is often used by industry to coat new or resharpened saw teeth, router bits, drills, end mills, chisels, tapes, and other cutting tools. Cutting tool edges are coated before the tools are packaged and shipped. This coating prevents the cutting edge from rusting and dulling during shipping and storage. Gears, hose fittings, and machine parts, such as lathe chucks, are often cold-dip coated prior to packaging. The coating also absorbs friction, shock, and seals out fingerprints.

Another type of cold dip substance is a vinyl-base, solvent-type liquid material. Vinyl-base material is generally not heated and it is applied by the dipping, brushing, or spraying technique. This material is often used to protect metal, ceramic, glass, and other surfaces too large to be placed in a heated container of plastic.

Vinyl-base "cold" material is used as a coating to protect product surfaces from scratches, rust, dirt, corrosion, and abrasion during storage, shipping, and fabrication. Often, the material is used as a masking for product protection during finishing, installation, or construction. An example would be a vinyl-base "mask" used on a plumbing fixture or a piece of machinery to prevent an area from being plated during a plating operation. Another "mask" may then

242

be used over the plated area to protect the surfaces from rough handling during installation.

REVIEW

1. Explain how CAB may be recycled.

2. Name the plastic family of which CAB is a member.

3. List four reasons for coating a product with peelable butyrate.

4. Name products other than a woodworking chisel edge which may be coated with CAB.

5. Name a material other than colorants and plastic found in the peelable butyrate.

6. Explain how to prevent CAB from overheating and burning in the dip tank.

7. Name two cutting tool edge cleaning techniques.

SELECTED BIBLIOGRAPHY

Agranoff, Joan, ed. *Modern Plastics Encyclopedia.* New York: McGraw-Hill Book Company, 1976-77.

Baird, Ronald J. *Industrial Plastics.* South Holland, Illinois: The Goodheart-Willcox Company, Inc., 1971.

Milby, Robert V. *Plastics Technology.* New York: McGraw-Hill Book Company, 1973.

Richardson, Terry A. *Modern Industrial Plastics.* Indianapolis: Howard W. Sams & Company, Inc., 1974.

Rosato, Dominick V. ed. *Plastics Industry Safety Handbook.* Boston: Cahners Books, 1973.

Cold Dip Coating (Peelable Butyrate)

EQUIPMENT AND MATERIAL SUPPLIERS

1. Dip Seal Plastics, Inc., 2311 23rd Avenue, Rockford, Illinois 61101.

2. Evans Manufacturing Inc., 4049 Otis, Warren, Michigan 48091.

3. Lacquer & Chemical Corporation, 214 40th Street, Brooklyn, New York 11232.

4. Thermo-Cote, Inc., P.O. Box 300, 267 Vreeland Avenue, Patterson, New Jersey 07513.

5. Western Coating Company, Box 598, Oakridge Station, Royal Oak, Michigan 48073.

A deep fryer can be purchased at any appliance store. Melt tanks and peelable butyrate are available from suppliers 1, 2, 4, and 5. Vinyl-base coatings can be purchased from suppliers 3 and 4.

Section VIII: Coatings

Assignment 26
Hot Dip Coating
(Vinyl Dispersion)

OBJECTIVES

To prepare a mold for a hot dip coating.

To coat a mold with a hot dip vinyl dispersion.

INTRODUCTION

The hot dip coating process is used by industry to make hollow one piece products. It is called "hot dip coating" because the mold is hot when dipped into the coating. The coating and mold are then cooled. The coating (product) may be peeled from the mold.

Products made by this process include toys, gloves, comb cases, spark plug covers, and automotive pump diaphragms. Hot dip coating is also used for protective coatings on certain metal products. An industrial, batch-type hot dip coating machine used to make plastic gloves is shown in Fig. 26-1.

Molds used for hot dipping are generally made of aluminum. They are called "mandrels" or "plugs." Aluminum is only one material that can be coated with a hot dip vinyl dispersion. Glass and ceramic materials can also be hot dip coated.

A plastic used as a coating for the hot dip process is vinyl. Before the vinyl is sold, the manufacturer often adds a plasticizer to it. The plasticizer makes the finished product flexible. Without the plasticizer, vinyl is rigid and stiff. the more plasticizer added (within limits), the more flexible the vinyl.

Fig. 26-1 Batch-type hot dip coating machine.
(Courtesy McNeil Akron Division/McNeil Corporation)

SAFETY

The following precautions should be taken when hot dip coating:

1. Work in a well-ventilated area. Do not breathe the methylene chloride fumes. Do not breathe the hot vinyl dispersion fumes during the curing stage.

2. Work on a heat-resistant surface.

3. Wear safety glasses and heat-resistant gloves.

4. Keep a general purpose fire extinguisher in the work area.

5. Learn the safe operation of the oven.

6. Do not heat the vinyl dispersion with an open flame.

7. Do not heat the vinyl dispersion above 350° F. The material will decompose and burn above this temperature.

8. Handle the hot mold with pliers.

9. Do not cut yourself with the sharp utility knife.

EQUIPMENT AND MATERIALS

The equipment and materials needed for hot dip coating are:

1. Safety glasses.
2. Heat-resistant gloves.
3. Heat-resistant surface.
4. Oven (vented industrial or laboratory type unit containing a heat circulation fan).
5. Mold or mandrel.
6. Wire hook (shaped from light gauge steel wire).
7. No. 65 hot dip vinyl dispersion plastic.
8. Methylene chloride.

Fig. 26-2 Hot dip coating equipment and materials.

Fig. 26-3 Cleaning mold with steel wool and methylene chloride.

Fig. 26-4 Attaching hook and mold to upper oven rack.

9. Board.
10. Steel wool.
11. Pliers.
12. Utility knife.
13. Clean cloth or paper towels.

HOT DIP COATING LAB PROCEDURE

1. Obtain all the items listed under "Equipment and Materials" needed to make the trailer hitch ball cover in this assignment. See Fig. 26-2.

2. Clean the trailer hitch ball cover mold with steel wool and methylene chloride. This removes dirt and scale. See Fig. 26-3. Follow the cleaning procedure described in Assignment 25 for cleaning items that were cold dip coated. Do not breathe the methylene chloride fumes.

3. Let the mold surface dry.

4. Place the wire hook through the mold stud hole.

5. Attach the hook and mold to the upper oven rack. See Fig. 26-4.

6. Heat the oven and mold by setting the oven temperature switch at 350° F. Let the mold heat for 15 to 20 minutes. The oven temperature and heating time for each plastic is different. Follow your instructor's or the plastic manufacturer's recommendations for oven temperature settings and mold preheating times.

7. Wear heat-resistant gloves during Steps 7-12. Remove the mold from the oven by gripping it with pliers. Be sure to wear heat-resistant gloves and use pliers whenever touching or moving the hot mold.)

8. Dip the hot mold completely into the hot dip vinyl dispersion. See Fig. 26-5. Ask your instructor how long to keep the mold in the vinyl.

Hot Dip Coating (Vinyl Dispersion)

Fig. 26-5   Dipping the hot mold.

Fig. 26-6   Attaching coated mold to upper oven rack.

Fig. 26-7   Cooling mold.

Note: the longer the dip time and the higher the mold temperature, the thicker the vinyl coating.

9. Remove the mold from the hot dip vinyl and let the plastic drip back into the container.

10. Hook the coated mold to the upper oven. See Fig. 26-6.

11. Heat (cure) the coated mold for 15 to 20 minutes at a 350° F. oven temperature. Do *not* touch the coated surface. The coating is cured when its surface changes from a glossy to dull and back to a glossy finish. If a thicker coating is needed, repeat Steps 7, 8, 9, 10 and 11. Oven temperatures needed for each plastic vary. Follow your instructor's or plastic manufacturer's recommendations for oven temperature settings and heating times.

12. Cool the cured, coated mold in cold water for 2 to 5 minutes. See Fig. 26-7.

13. Dry the coating and carefully slip it off the mold. See Fig. 26-8.

14. Check the thickness of the coating. If the coating is too thin, heat the mold longer or at a higher temperature. **Be sure to set the oven temperature high enough.** Not dipping the mold long enough, or cooling it too much before coating, may also make a thin coating. If the mold coating is too thick, reduce the preheating temperature, or dipping time.

15. Check the coating for burns. Burns are caused by post-heating the mold too long and at too high a temperature.

16. Trim the bottom edge of the trailer hitch ball cover with a utility knife. See Fig. 26-9. *Be careful not to cut yourself.*

Section VIII: Coatings

Fig. 26-8   Slipping the coating off the mold.

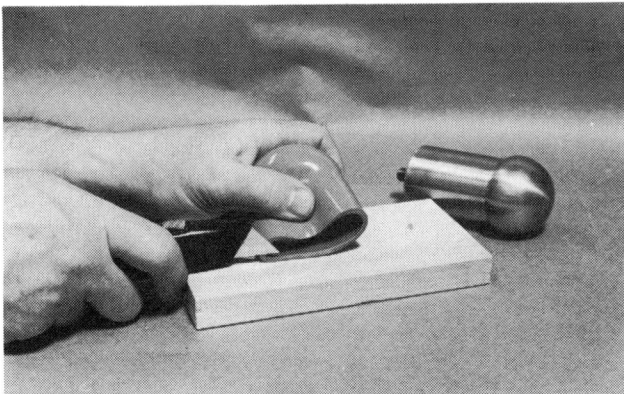

Fig. 26-9   Trimming bottom edge.

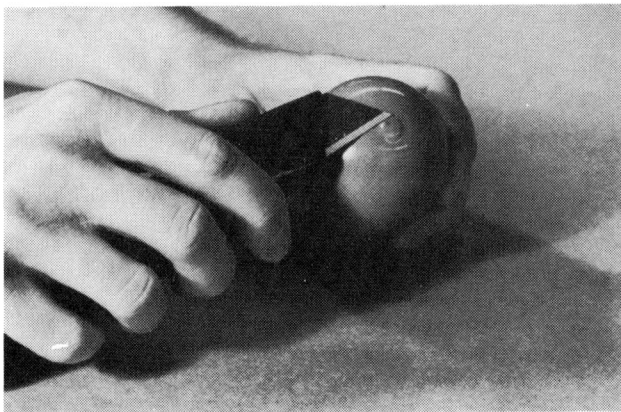

Fig. 26-10   Trimming top surface.

17. Trim the "drip" from the top surface of the finished cover with a utility knife. See Fig. 26-10. *Be careful not to cut yourself.*

CONCLUSION

Other products that were coated or made by hot dip (vinyl dispersion) coating are shown in Fig. 26-11. Use the procedure just described to produce these items.

Dipping a mold into the vinyl dispersion is one way to coat a product. There are other coating methods, too. For example, the dispersion can be sprayed onto the inside of railroad chemical tank car walls. The cured dispersion gives the car a protective lining.

Some products, such as carpets, have a soft vinyl dispersion backing. The dispersion is placed on the back of the carpet with a plastic foaming process.

REVIEW

1. List four products produced by the hot dip coating process.

2. Name the type of plastic used for the hot dip coating on your project.

3. Explain why plasticizers are added to some plastics.

4. Name the mold material used for your hot dip project.

5. List two methods for placing a vinyl dispersion on a product other than dipping.

6. Describe how to correct a mold that has a thin coating.

7. Describe how to correct a mold that has a burned coating.

Hot Dip Coating (Vinyl Dispersion)

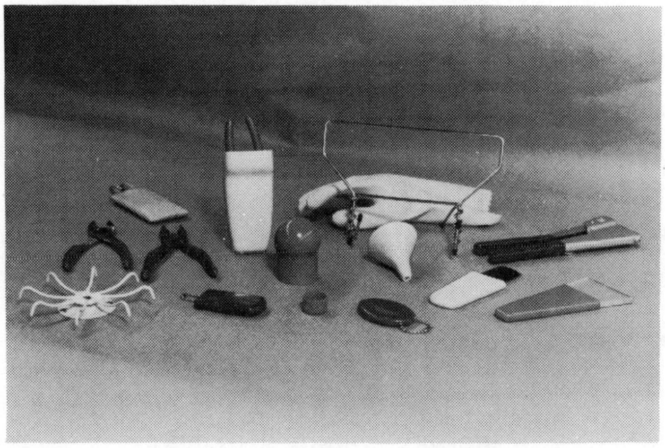

8. Explain why a mold can have a rough or crumbly coating.

SELECTED BIBLIOGRAPHY

Agranoff, Joan, ed. *Modern Plastics Encyclopedia.* New York: McGraw-Hill Book Company, 1976-77.

Baird, Ronald J. *Industrial Plastics.* South Holland, Illinois: The Goodheart-Willcox Company, Inc., 1971.

Coatings—Teacher's Manual. Indianapolis: Howard W. Sams & Company, Inc., 1974.

Milby, Robert V. *Plastics Technology.* New York: McGraw-Hill Book Company, 1973.

Richardson, Terry A. *Modern Industrial Plastics.* Indianapolis: Howard W. Sams & Company, Inc., 1974.

Rosato, Dominick V., ed. *Plastics Industry Safety Handbook.* Boston: Cahners Books, 1973.

EQUIPMENT AND MATERIAL SUPPLIERS

1. Brodhead-Garrett, 4560 East 71st Street, Cleveland, Ohio 44105.

2. Cope Plastics, Inc., 4441 Industrial Drive, Godfrey, Illinois 62035.

3. Delvie's Plastics, Inc., 2320 South West Temple, P.O. Box 1415, Salt Lake City, Utah 84110.

4. Graves-Humphreys, Inc., 1948 Franklin Road, P.O. Box 1347, Roanoke, Virginia 24033.

5. Industrial Arts Supply Company, 5724 West 36th Street, Minneapolis, Minnesota 55416.

Fig. 26-11   Examples of hot dip coated products.

6. McKilligan Industrial Supply Corporation, 494 Chenango Street, Binghamton, New York 13901.

7. Paxton/Patterson, 5719 West 65th Street, Chicago, Illinois 60638.

Ovens can be purchased from any of the suppliers listed above. Vinyl dispersions are available from suppliers 1, 2, 3, 4, 5, and 7. Vinyl dispersion molds are available from suppliers 1, 2, 3, 5, 6, and 7.

Hot Dip Coating (Vinyl Dispersion)

Assignment 27
Fluidized Bed Coating

OBJECTIVES

To prepare a surface for a fluidized bed coating.

To fluidized bed coat a product.

INTRODUCTION

Fluidized bed coating is a process for coating products. The coating is done in a tank that can air or nitrogen-float plastic powder. First, the product is heated above the melting point of the plastic powder. The heated product is then dipped in the powder floating in the tank. After coating, the product is heated to completely fuse the coating.

This process is used by industry to coat dishwasher racks, refrigerator shelves, marine hardware, outdoor furniture, transformer tanks, food handling equipment, and outdoor electrical equipment. Products are often fluidized bed coated instead of painted. For example, many small tractor and lawn mower body parts are fluidized bed coated.

Carbon steel, stainless steel, cast iron, aluminum, zinc alloy, copper, bronze, and brass are all metals that can be fluidized bed coated. Glass and ceramics can also be fluidized bed coated. Before these materials are coated, they must be cleaned and dried. Make sure to clean off any oil film.

A long lasting fluidized bed coating can be placed on parts. The part surface must be cleaned and coated or primed with a special adhesive. The part is then ready to be fluidized bed coated.

Section VIII: Coatings

SAFETY

The following precautions should be taken when fluidized bed coating:

1. Work in a well-ventilated area. Do not breathe the methylene chloride fumes. Do not breathe the hot plastic coating fumes during the curing stage.

2. Work on a heat-resistant surface.

3. Wear safety glasses and heat-resistant gloves.

4. Keep a general purpose fire extinguisher in the work area.

5. Learn the safe operation of the oven.

6. Handle the hot part to be coated with pliers.

7. Be careful not to cut yourself with the sharp utility knife.

8. Do not place your hands and fingers in the plastic powder.

EQUIPMENT AND MATERIALS

The equipment and materials needed for fluidized bed coating are:

1. Safety glasses.
2. Heat-resistant gloves.
3. Heat-resistant surface.
4. Oven (vented industrial or laboratory unit containing a heat circulating fan).
5. Fluidized bed tank (charged with plastic coating powder).
6. Article to be coated.
7. Wire hook (shaped from light gauge steel wire).
8. Regulated air supply.
9. Air hose.
10. Methylene chloride.
11. Steel wool.
12. Utility knife.
13. Clean cloth or paper towels.
14. Pliers

Fig. 27-1 Fluidized bed coating equipment and materials.

Fig. 27-2 Cleaning with steel wool and methylene chloride.

Fig. 27-3 Attaching hook and pliers to oven rack.

FLUIDIZED BED COATING LAB PROCEDURE

1. Obtain all the items listed under "Equipment and Materials" needed to fluidize bed coat plier handles. See Fig. 27-1.

2. Clean the handles with steel wool and methylene chloride. This removes dirt, scale, and rust. See Fig. 27-2. Follow the same cleaning procedure illustrated in Assignment 25 for cleaning items that were cold dipped. Do not breathe the methylene chloride fumes.

3. Let the plier handles dry.

4. Place the wire hook through the slip joint hole of the pliers.

5. Attach the hook and pliers to the oven rack. See Fig. 27-3.

6. Set the oven temperature switch to 350° F. and heat the pliers. Let the pliers heat up for 15 to 20 minutes. (The oven temperature and heating time for each plastic is different. Follow your instructor's or plastic manufacturer's recommendations for temperature and time.)

7. Make sure that the air supply valve is turned off at the tank or at the air line regulator. Connect the air line or hose to the tank. See Fig. 27-4.

Fig. 27-4 Connecting air hose to tank.

Fig. 27-5 Turning air valve.

Fig. 27-6 Moving pliers in fluidized bed.

Fig. 27-7 Hooking vinyl coated pliers to upper oven rack.

8. Slowly turn on the air pressure valve at the tank or at the air line regulator. Keep turning the air valve until the plastic powder in the partially filled tank bubbles or flows. See Fig. 27-5.

9. While wearing heat-resistant gloves, remove the pliers from the oven using a second set of pliers. *Always wear the heat-resistant gloves and use pliers to touch or move the hot pliers.*

10. Move the hot pliers through the fluidized bed. Use a circular motion and move the pliers around for 15 to 20 seconds. See Fig. 27-6. Place **only** the part of the pliers to be coated in the plastic.

11. Hook the vinyl coated pliers to the upper oven rack. See Fig. 27-7.

12. Heat (cure) the coated pliers for 15 to 20 minutes at 350° F. oven temperature. Do not bump the coated tool surface. The coating is cured when the surface changes from a fuzzy, dull finish to a smooth, shiny color. Oven temperature for each plastic will vary. Follow your instructor's or plastic manufacturer's recommendations for oven temperature settings and times.

13. Cool the pliers in cold water for 2 to 5 minutes. See Fig. 27-8.

Fig. 27-8 Cooling pliers with cold water.

Fluidized Bed Coating

14. Dry the pliers.

15. Check the thickness of the coating. If it is too thin, heat the part longer or at a higher temperature. **Be sure to set the oven temperature high enough.** Not dipping the part long enough, or cooling it too much before dipping, may also make a thin coating. If the coating is too thick, reduce the heating or dipping time. Also, the part may not have moved enough in the powder.

16. Check the part for uncoated spots or pinholes. This defect may occur if the part is too big to be dipped into the tank, the coating is too thin, or the part is not moved enough in the powder.

17. Check the part for an edge or spot showing through the coating. This can occur if the part is pre-heated too long, post-heated too long, has too thick a coating, or has sharp edges that were not removed.

18. Check the coating for burns. Burns are caused by post-heating or pre-heating too long.

19. Check the coating for roughness. If the part coating is rough, the part may not have been post-heated or pre-heated long enough.

20. Check the coating to see if it sticks to the primered part. If not, the part may have been pre-heated too long. It may also need cleaning.

21. Trim the excess plastic from each plier handle with a utility knife. See Fig. 27-9. Be careful not to cut yourself. Some plastic coatings cure to a very hard finish and can not be trimmed. These coatings must be trimmed with a utility knife right after they have been fluidized bed coated. Follow your instructor's recommendations for trimming the excess plastic.

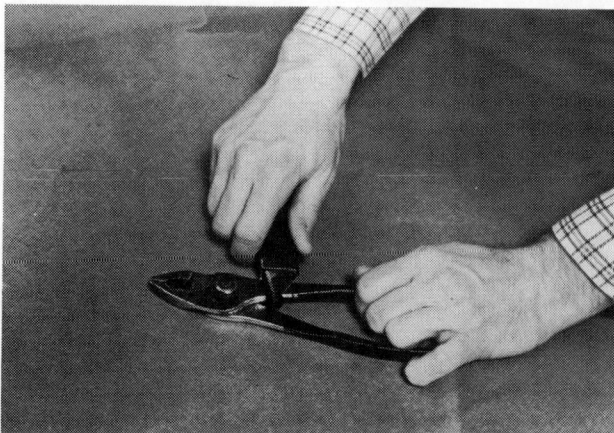

Fig. 27-9   Trimming excess plastic from pliers.

22. The finished plier handle coating is shown in Fig. 27-10. The coating should be glossy, smooth, and free of pinholes.

CONCLUSION

Shown in Fig. 27-11 are many products that were fluidized bed coated. The procedure just described may be used to coat these items.

Many types of plastics may be used to fluidize bed coat. Vinyl is often used in school shops. Cellulosic, epoxy, nylon, polyethylene, chlorinated polyether, and polyester plastics are also used for fluidized bed coating. Some of these plastics belong to the thermoplastic family and others belong to the thermoset family.

REVIEW

1. List four products coated by the fluidized bed process.

2. Name three plastics that may be used in the fluidized bed coating process.

3. Define a *thermoset plastic*.

4. Name a gas that floats the plastic in the fluidized bed tank.

5. Explain why the product must be preheated.

6. Explain why the product must be post-heated.

7. Describe how to correct a part that has a pinhole coating.

8. Describe how a part can be too thinly coated.

9. Describe how to correct a part that has a burnt coating.

10. List all the steps in fluidized bed coating a product.

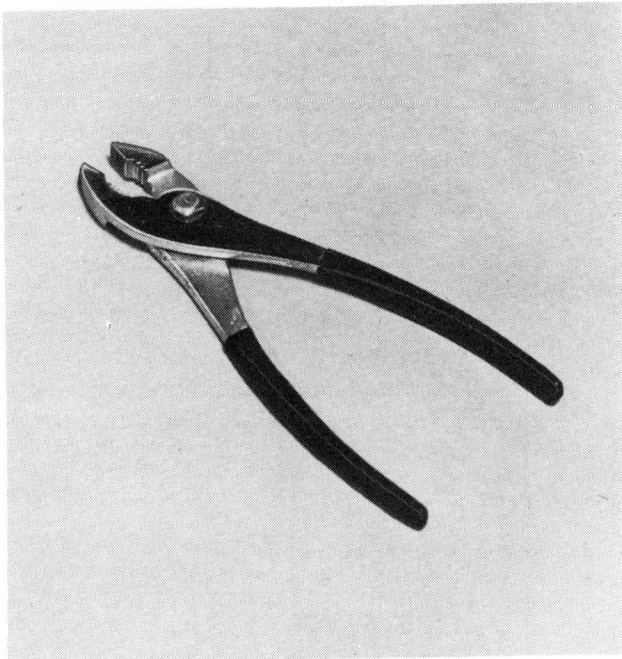

Fig. 27-10   Finished coating.

Fig. 27-11   Examples of fluidized bed coated products.

Fluidized Bed Coating

SELECTED BIBLIOGRAPHY

Agranoff, Joan, ed. *Modern Plastics Encyclopedia.* New York: McGraw-Hill Book Company, 1976-77.

Baird, Ronald J. *Industrial Plastics.* South Holland, Illinois: The Goodheart-Willcox Company, Inc., 1971.

Coatings—Teacher's Manual. Indianapolis: Howard W. Sams & Company, Inc., 1974.

Milby, Robert V. *Plastics Technology.* New York: McGraw-Hill Book Company, 1973.

Patton, William J. *Plastics Technology: Theory, Design, and Manufacture.* Reston, Virginia: Reston Publishing Company, Inc., 1976.

Richardson, Terry A. *Modern Industrial Plastics.* Indianapolis: Howard W. Sams & Company, Inc., 1974.

EQUIPMENT AND MATERIAL SUPPLIERS

1. Brodhead-Garrett, 4560 East 71st Street, Cleveland, Ohio 44105.

2. Delvie's Plastics, Inc., 2320 South West Temple, P.O. Box 1415, Salt Lake City, Utah 84110.

3. Industrial Arts Supply Company, 5724 W. 36th St., Minneapolis, Minnesota 55408.

4. McKilligan Industrial Supply Corporation, 494 Chenango Street, Binghamton, New York 13901.

5. Polymer Corporation, 2120 Fairmont Avenue, Reading, Pennsylvania 19603

Fluidized bed tanks and plastics for fluidized bed coating can be purchased from any of the suppliers listed above. Ovens are available from suppliers 1, 2, 3, and 4.

Section IX
Decorating

Assignment 28
Hot Foil Stamping

OBJECTIVES

To select a hot stamp die and roll of foil.

To adjust the stamping equipment.

To make a hot stamped product.

INTRODUCTION

Hot foil stamping was one of the first methods used to decorate plastic. A heated die is used to transfer a design from a dry decorative film to the part being decorated. This process is often called "dry printing." The process can be automatic, semiautomatic, or manual. In Fig. 28-1 a large, industrial hot foil stamping machine is shown. It is stamping lines on ping pong tables.

The hot foil stamping process works as follows. A die made of silicone, brass or steel is heated. A decorative foil is passed under the die. Then, the die is lowered and held against the foil.

The heated die is pressed **firmly** against the foil. This forces the foil onto the part. This process releases, transfers, and fuses a decorative ink or paint from the film to the part. Then, the die retracts.

The foil cools on the plastic. Then, the foil not touched by the die is stripped from the part. This leaves only the design of the die, foil-stamped on the part. The stamped parts can be handled immediately.

There are a number of hot stamp films used. Some films leave the product with a metallic, colored, wood-grained, marbled, reptile, or mother-of-pearl finish.

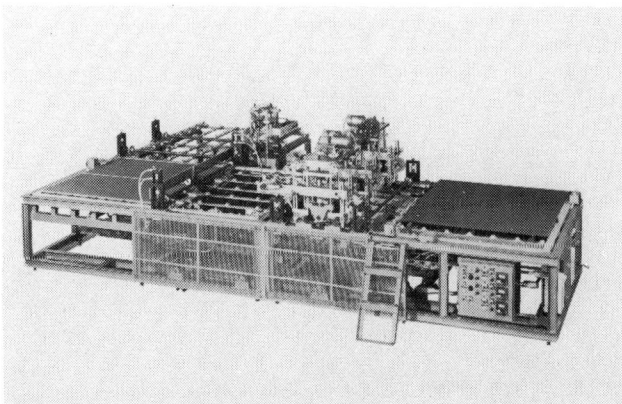

Fig. 28-1   Industrial hot foil stamping machine. (Courtesy Dri-Print Foils, Inc.)

Section IX: Decorating

Foils are made up of various layers of material. Each reacts in different ways under die heat and pressure. A few of the foil layers include carrier layers, release layers, lacquer layers, decorative layers, and adhesive layers. Each layer serves a different function during hot foil stamping.

SAFETY

The following precautions should be taken when hot foil stamping:

1. Wear safety glasses.

2. Learn the safe operation of the hot stamping equipment.

3. Do not burn yourself on the hot press parts.

4. Do not pinch your hands and fingers between the die and press table.

5. Make sure all machine control and moving part guards are in place and working.

6. Use the two hand safety lock switch system.

EQUIPMENT AND MATERIALS

The equipment and materials needed for hot foil stamping are:

1. Safety glasses.
2. Hot stamping machine.
3. Hot stamping film.
4. Part to be stamped.
5. Air supply (adjustable high pressure).
6. Hot stamping die or type.
7. Chase.
8. Part fixture.

HOT FOIL STAMPING LAB PROCEDURE

A school laboratory hot foil stamping machine is illustrated in

Hot Foil Stamping

Fig. 28-2 School laboratory hot foil stamping machine.

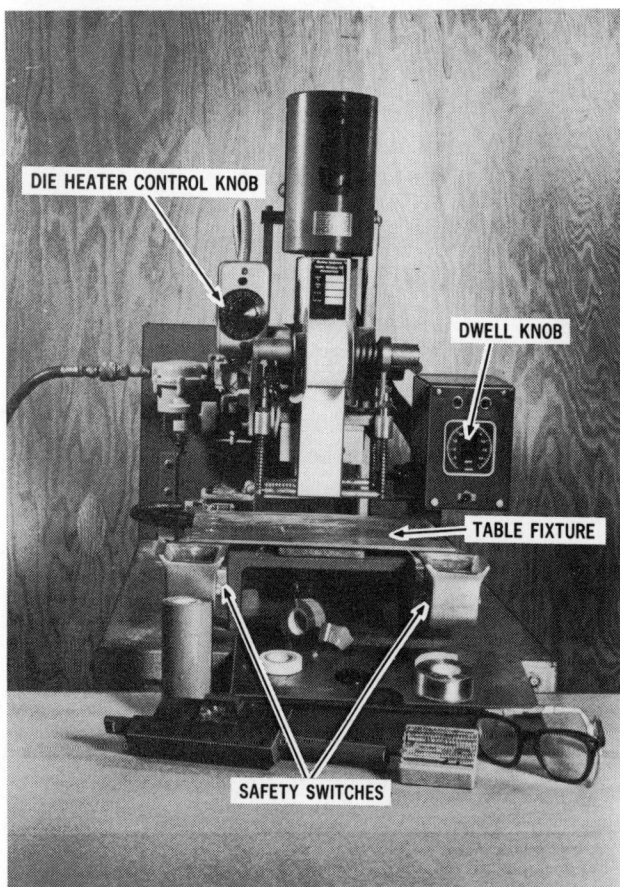

DIE HEATER CONTROL KNOB

DWELL KNOB

TABLE FIXTURE

SAFETY SWITCHES

Fig. 28-3 Hot foil stamping equipment and materials.

Fig. 28-2. Your school shop may have similar equipment. Be sure you understand how the equipment works before you use it.

1. Obtain all the items listed under "Equipment and Materials" needed to hot foil stamp a product. See Fig. 28-3.

2. Turn on the die heater switch.

3. Adjust the die heater control knob to the proper temperature. See Fig. 28-4. Each type of hot stamp foil requires a different temperature. Ask your instructor what temperature to use. A temperature of 325° F. was used in this assignment. Make sure the foil is the correct one for the plastic to be stamped.

4. Let the die heat for 15 to 20 minutes.

5. Turn on the air pressure to the die head air cylinder. See Fig. 28-5. Air pressure controls the force at which the heated die strikes the foil and the plastic product. Generally, the harder the plastic, the higher the air pressure used. Ask your instructor what setting to use. 40 psi was used in this assignment.

6. Adjust the dwell knob to the correct setting. (*Dwell* is a measure of time.) See Fig. 28-6. Ask your instructor what dwell setting to use. A setting of 0.5 seconds was used in this assignment. (In this assignment, the dwell time is the time the hot die touches the foil and plastic product.)

7. Place the product to be stamped in the table fixture. See Fig. 28-7. (A fixture is a holder that lines up the product with the die. It makes sure that the printing is done on the center of the product.) Often the fixture will have a flexible pad on which the product is placed. The pad flexes a little when the die

264

Fig. 28-4 Adjusting die heater control knob.

Fig. 28-5 Turning on air pressure.

Fig. 28-6 Adjusting dwell knob.

stamps the product. This flexing lets the die stamp the product evenly.

8. Press **both** hand safety switches **at the same time.** See Fig. 28-8. Turning on the safety switches allows air to fill the die head air cylinder. This causes the die to strike the foil and product.

9. Remove the stamped product after the die returns to its original position.

10. Look at the stamped area of the product. See Fig. 28-9. The stamping should be even and clear with no indentations. The hot stamped insert is ready to be solvent-cemented to the name tag, shown in Assignment 3 (injection molding).

Fig. 28-7 Placing product in table fixture.

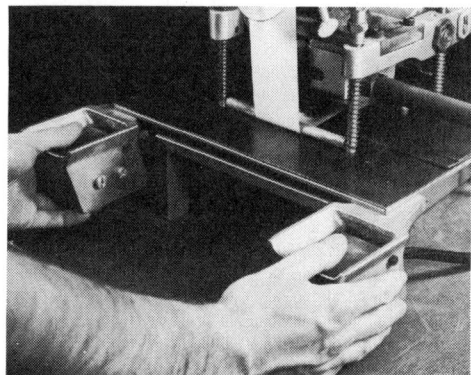
Fig. 28-8 Pressing both hand safety switches.

Hot Foil Stamping

CONCLUSION

Many hot foil stamped products are shown in Fig. 28-10. Each is made by using a different die with the procedure just described.

The process just described is slow. There are ways to speed up the process. Industry uses hot stamping machines for fast, mass production decorating. By using certain machines, dies, and foils, industry can hot stamp 60 to 4,600 pieces per hour.

Most dies used for long production runs are made of steel. Brass dies are used for medium pressure and short production runs.

Many products, such as hot stamped woodgrained TV cabinets, receive a continual or solid textured design. A heated solid or textured design silicone die is used for this purpose.

Often a product needs foil decoration placed on a flat surface or a raised area. A hand-held, portable, heated, silicone hot stamp roller is used. Foil is placed on the flat or raised product surface. The roller is heated. Then, it is slowly rolled over the foil. This causes the decorative layer to transfer and decorate the surface.

Many of the thermoplastics may be hot foil stamped. Some of these plastics include vinyls, styrenes, acrylics, and olifins. Often the olifins must be surface heat-treated before hot stamping. Due to their brittleness, thermosets are not often hot foil stamped.

REVIEW

1. Name the type of plastic you hot foil stamped.

2. Give another name for dry printing.

Fig. 28-9   Hot foil stamped insert cemented to name tag.

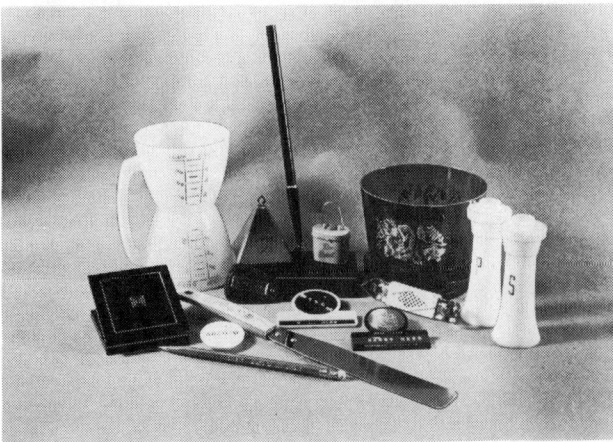

Fig. 28-10   Examples of hot foil stamped products.

3. List three products decorated by hot foil stamping.

4. Name the material used to make the hot stamp die.

5. Describe each step in the hot foil stamping procedure.

6. Show two hot foil stamped products having different stamped designs.

7. Explain the reason for a dwell setting.

8. Describe what to look for in a perfectly hot foil stamped product.

9. Explain the reason for the hot stamp table fixture.

10. Describe how to adjust the machine air pressure for hot stamping soft plastics.

11. Name two thermoplastics commonly hot foil stamped.

12. Name the hot stamping die material used for long production stamping runs.

SELECTED BIBLIOGRAPHY

Agranoff, Joan, ed. *Modern Plastics Encyclopedia.* New York: McGraw-Hill Book Company, 1976-77.

Baird, Ronald J. *Industrial Plastics.* South Holland, Illinois: The Goodheart-Willcox Company, Inc., 1971.

Milby, Robert V. *Plastics Technology.* New York: McGraw-Hill Book Company, 1973.

Richardson, Terry A. *Modern Industrial Plastics.* Indianapolis: Howard W. Sams & Company, Inc., 1974.

EQUIPMENT AND MATERIAL SUPPLIERS

1. Brodhead-Garrett, 4560 East 71st Street, Cleveland, Ohio 44105.

2. Graves-Humphreys, Inc., 1948 Franklin Road, P.O. Box 1347, Roanoke, Virginia 24033.

3. McKilligan Industrial Supply Corporation, 494 Chenango Street, Binghamton, New York 13901.

The hot stamping machine, tape and type may be purchased from any of the suppliers listed above.

Assignment 29
Machine Engraving

OBJECTIVES

To select the proper type, cutter, and plastic for engraving.

To set up the engraving machine.

To machine engrave a product.

INTRODUCTION

Machine engraving is a low volume process used to cut lines or designs into a plastic surface. It can be done by automatic or manual means.

Machine engraving is one method of permanently decorating or marking plastic. Some items engraved include signs, dials, tags, plaques, instrument panels, house numbers, name plates, molds, patterns, and dies. Fig. 29-1 shows a medium-size industrial engraving machine.

Machine engraving works as follows. A stylus is made to follow the grooves in a die. The stylus is fastened to a series of arms (called a pantograph). The arms are attached to a high-speed, rotating cutter. As the stylus follows the die, its movements are transferred to the cutter. The cutter traces out the same design on the sheet of plastic.

The pantograph is usually adjusted to make the cut design smaller than the original die. After the adjustments are made, the cutter is turned on and the stylus is used to trace the die. The cutter engraves the plastic as it follows the movements of the stylus. Finally, the die design is cut into the plastic to decorate or identify the product.

Fig. 29-1   Industrial engraving machine. (Courtesy H.P. Preis Engraving Machine Company)

Machine Engraving

SAFETY

The following precautions should be taken when machine engraving:

1. Wear safety glasses.

2. Learn the safe operation of the engraving machine.

3. Make sure all machine control and moving part guards are in place and working.

4. Keep your hands and fingers out of the cutting area.

5. Make sure the cutter is tightly fastened in the chuck before turning on the machine.

6. Do not cut your hands or fingers while tightening the cutter in the machine chuck.

7. Keep your hands and fingers away from the drive belts.

8. Be careful not to cut your fingers or hands with the stylus.

9. Make sure the machine is unplugged when making all adjustments.

EQUIPMENT AND MATERIALS

The equipment and materials needed for machine engraving are:

1. Safety glasses.
2. Engraving machine.
3. Type.
4. Cutter (diamond point, carbide-tipped, high speed steel, or tipped-off point).
5. Plastic.
6. Stylus.
7. Type holder.
8. 2 plastic clamps.
9. 2 type clamps.

Fig. 29-2 School laboratory engraving machine.

Fig. 29-3 Machine engraving equipment and materials.

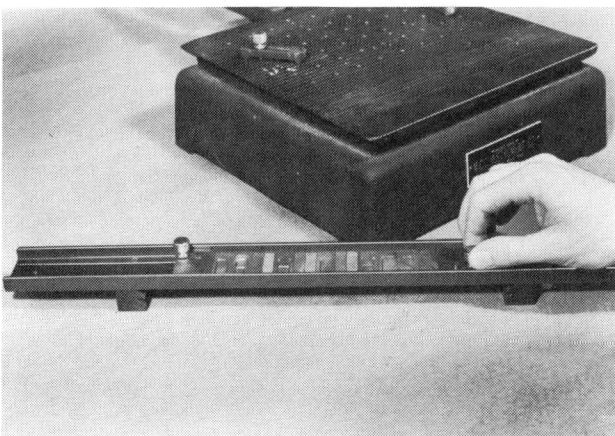

Fig. 29-4 Clamping type in holder.

MACHINE ENGRAVING LAB PROCEDURE

Shown in Fig. 29-2 is a small, school laboratory engraving machine. Your shop may have a similar one. Be sure you understand how to operate the machine before starting this assignment.

1. Obtain all the items listed under "Equipment and Materials" needed to engrave a plastic product. See Fig. 29-3.

2. Select the type style.

3. Center the total copy in the type holder. Make sure the spelling, punctuation, and spacing are correct.

4. Push the pieces of type tightly together.

5. Clamp the line of type tightly into the type holder. See Fig. 29-4.

6. Select the proper cutter.

7. Place the cutter in the engraving machine chuck, as in Fig. 29-5. Ask your instructor to show you which cutter to use. Be sure the cutter is tight in the chuck. (The cutter tip shape will form the shape of the engraved line. Cutter tips are made of high speed steel, carbide inserts, or diamond ends. Cutter tip shapes are vee-pointed or flat tipped for most plastic engraving.)

8. Adjust the engraver ratio by matching the ratio holes (lettered or numbered) on each arm. Ask your instructor which letters or numbers to use.

9. Place the pin through the matched holes on each arm. See Fig. 29-6. This controls the die-to-engraved design-height ratio.

Machine Engraving

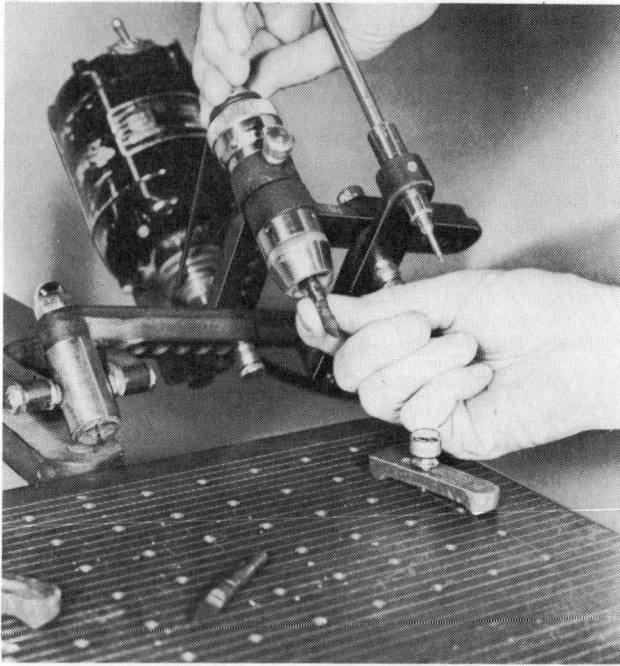

Fig. 29-5 Placing cutter in machine chuck.

Fig. 29-6 Inserting pin.

Fig. 29-7 Sliding type holder onto engraving table ways.

10. Tighten the locking knob for each pin. If the arms are adjusted right, they will form a parallelogram. The machine will engrave slanted letters if the arm holes are not matched.

11. Slide the type holder onto the engraver table ways. See Fig. 29-7. Make sure all type is within the range of the engraver arms (the pantograph). If the end type cannot be reached by the machine arms, slide the type holder toward the cutter.

12. Lock the type holder onto the table with set screws.

13. Locate the plastic by placing the stylus in the center of the line of type.

14. Move the plastic so its center is under the cutter. Make sure one straight edge of the plastic is parallel to the line of type.

15. Clamp the plastic to the engraver table. See Fig. 29-8.

16. Place the stylus in the bottom of the type groove.

17. Adjust the cutter so that it just rests on the plastic surface.

Fig. 29-8 Clamping plastic to table.

Section IX: Decorating

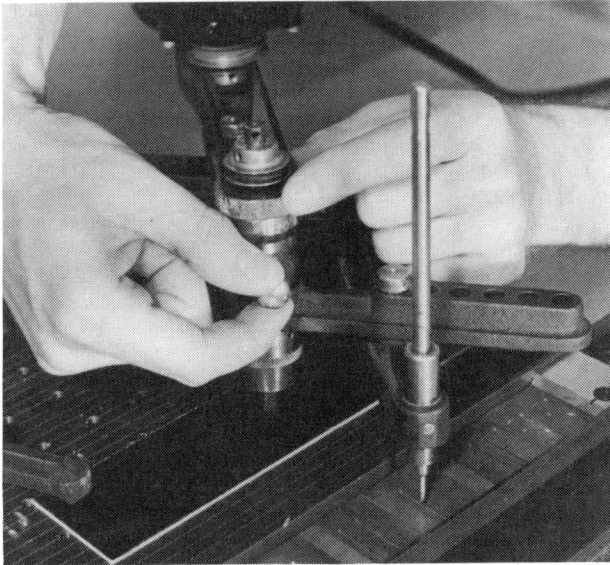

Fig. 29-9 Setting the depth control spindle.

Fig. 29-10 Engraving the first letter.

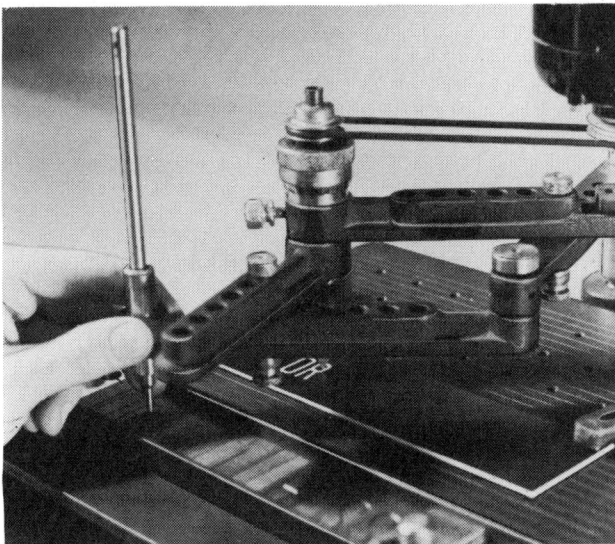

Fig. 29-11 Engraving the second letter.

18. Move the depth control spindle to the proper setting. See Fig. 29-9. This adjusts the depth of cut. Ask your instructor what depth setting to use for your project. Be sure to lock the depth spindle in place.

19. Turn the machine cutter on.

20. Engrave the first letter. See Fig. 29-10. Use a slow, steady motion. Engrave by starting at the top of the letter and working down.

21. Engrave the second letter with the same motions. See Fig. 29-11. Continue engraving all the letters in the same way. See Fig. 29-12. (Deep engraving often requires making many shallow cuts.)

22. Check the engraving. If the engraved area is too large or too small, the wrong engraver arm ratio numbers or letters were used. The engraved depth of cut must also be consistent. For an even cut, adjust the cutter to operate parallel to the engraver table. Slanted or curved engraved letters may be caused by not matching the proper holes on each engraver arm.

Fig. 29-12 Engraving the remaining letters.

Machine Engraving

Fig. 29-13   Finished engraving.

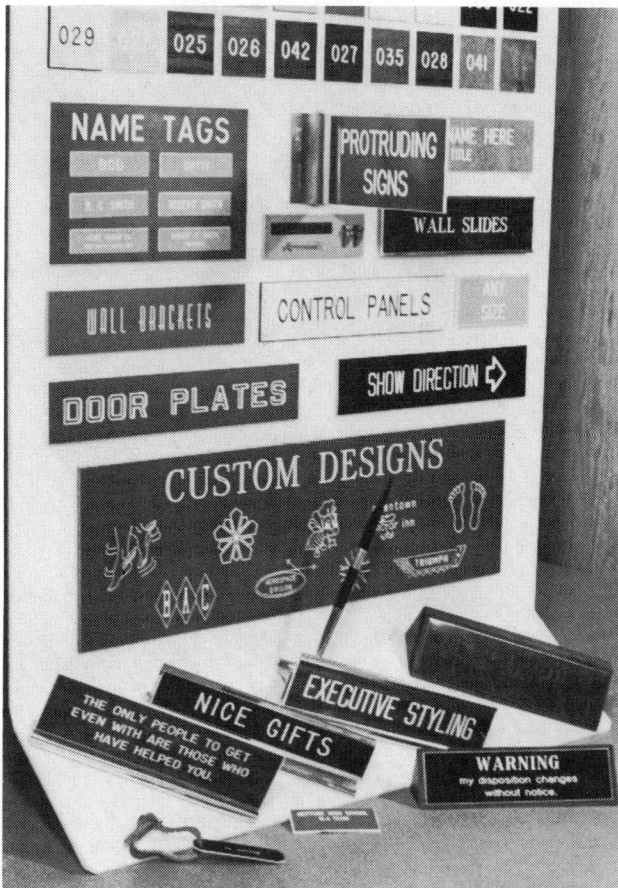

Fig. 29-14   Examples of machine engraved products.

The finished engraving, ready for display, is shown in Fig. 29-13. The engraved lines should have a consistent depth and width.

CONCLUSION

Figure 29-14 illustrates many engraved products. The procedure just described was used to engrave them. Laminated engraving plastic may be purchased in many colors, grain patterns, and designs.

When laminated (layered) plastics are engraved, the cutter tip removes the top plastic surface to expose the inside layer. This technique was just demonstrated.

Designs can be engraved into colored or textured plastic. When this plastic is engraved, the engraved area is filled with colored wax or painted to give contrast. Often, a clear plastic is engraved. The plastic is engraved on the back side with a reverse die. The engraved area is then filled with colored wax or painted.

The engraving machine can be made to engrave unusual products. A machine work table is often modified to engrave skis. The skis will have lettering or designs engraved on their surfaces. Also a stylus, work table, and cutter can often be modified to duplicate a three dimensional design in a block of plastic.

REVIEW

1. Name three plastics products machine engraved.

2. Describe each step of the machine engraving process.

3. Explain the reason for machine engraving plastics.

4. Describe the function of a pantograph.

5. Name the type of plastic that you machine engraved.

6. Name two materials used to make cutter tips.

7. List two common cutter tip shapes.

8. Describe how the pantograph arms are adjusted to get various die-to-engraved design-height ratios.

9. Describe how to adjust the cutter for the proper depth of cut.

10. Describe how to center the die with the plastic to be engraved.

SELECTED BIBLIOGRAPHY

Baird, Ronald J. *Industrial Plastics.* South Holland, Illinois: The Goodheart-Willcox Company, Inc., 1971.

Richardson, Terry A. *Modern Industrial Plastics.* Indianapolis: Howard W. Sams & Company, Inc., 1974.

EQUIPMENT AND MATERIAL SUPPLIERS

1. Brodhead-Garrett, 4560 East 71st Street, Cleveland, Ohio 44105.

2. Delvie's Plastics, Inc., 2320 South West Temple, P.O. Box 1415, Salt Lake City, Utah 84110.

3. Graves-Humphreys, Inc., P.O. Box 13407, 1948 Franklin Road, Roanoke, Virginia 24033.

4. McKilligan Industrial Supply Corporation, 494 Chenango Street, Binghamton, New York 13901.

The equipment and materials used in machine engraving can be purchased from any of the suppliers listed above.

Machine Engraving

Section X
Packaging

Assignment 30
Impulse Sealing

OBJECTIVES

To prepare impulse sealing equipment, plastic, and products.

To make an impulse sealed package.

INTRODUCTION

Impulse sealing is a heat sealing process. It is used in the packaging industry to protect products during shipping. Impulse sealing bonds a plastic to itself or to other plastics. Materials used include nylon, vinyl, polyethylene, polypropylene, and fluorocarbon films.

The impulse sealing process works as follows. The product to be sealed is placed between two pieces of plastic. Then, the edges of the two pieces of plastic are held together. They are placed between a movable and a stationary set of jaws in the sealing machine. The upper, or movable jaw, is then brought down on top of the two sheets of plastic.

The sheets are tightly pressed together against the stationary sealing jaw. When the two jaws make contact, a switch turns on an electrical element in the stationary jaw. Current flows in the element and heats it up. The machine timer is set to control how long the element heats.

The heat bonds the two sheets of plastic together to make the seal. When the heat switch shuts off, the plastic is left clamped until it cools. Then, the upper jaw is opened and the sealed plastic is removed.

SAFETY

The following precautions should be taken when impulse sealing:

1. Wear safety glasses.

2. Learn the safe operation of the impulse sealer.

3. Keep your hands and fingers away from the movable upper arm and the stationary sealing bar.

4. Learn how to connect and adjust the foot pedal of the impulse sealer.

5. Learn the proper impulse sealing procedure. This will help prevent damaging the machine and the package being sealed.

EQUIPMENT AND MATERIALS

The equipment and materials needed for impulse sealing are:

1. Safety glasses.
2. Impulse sealer.
3. Nylon, vinyl, polyethylene, polypropylene, or fluorocarbon film sealing plastic.
4. Sealing plastic feed mechanism.
5. Product to be packaged.
6. Scissors.

IMPULSE SEALING LAB PROCEDURE

A small, school laboratory impulse sealer is shown in Fig. 30-1. Plastic is stored and fed to the sealer from a separate mechanism.

1. Obtain all the items listed under "Equipment and Materials" needed to impulse seal a package. See Fig. 30-2.

2. Obtain the products to be impulse seal packaged. See Fig. 30-3.

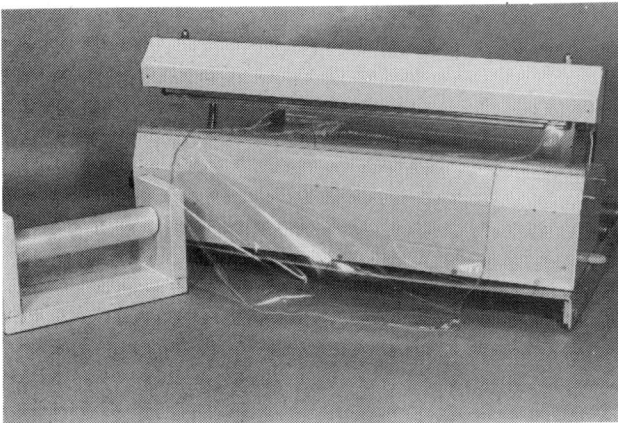

Fig. 30-1 School laboratory impulse sealer and feed mechanism.

Fig. 30-2 Impulse sealing equipment and materials.

3. Connect the foot pedal assembly to the impulse sealer. (Some sealers do not use a foot pedal.)

4. Attach a chain to the hook on the bottom of the machine. Make sure the chain passes through a hole in the bench.

5. Fasten the sealer to the bench with clamps or wood screws. Check with your instructor to determine how this should be done.

6. Lay the lower end of the foot pedal on the floor.

7. Hook the **upper** end of the foot pedal to the **bottom** of the chain. See Fig. 30-4. Make sure the pedal is fastened to the chain so that the sealer completely closes when the pedal is depressed.

8. Connect the electric cord to the sealer.

9. Adjust the sealing timer. See Fig. 30-5. This adjustment will be different for each type and thickness of plastic. Ask your instructor what setting to use with your equipment. (A setting of 2 seconds was used for this operation.) Generally, the thicker the plastic, the longer the sealing time.

10. Hold the partial plastic package between the upper jaw and the sealing bar (lower jaw). See Fig. 30-6.

11. Press the foot pedal fully until the buzzer sounds and then stops. (Some machines use a light in place of a buzzer.)

12. Hold the foot pedal down for the amount of time set on the timer. This will allow the sealed plastic to cool.

13. Place the product(s) in the partial package.

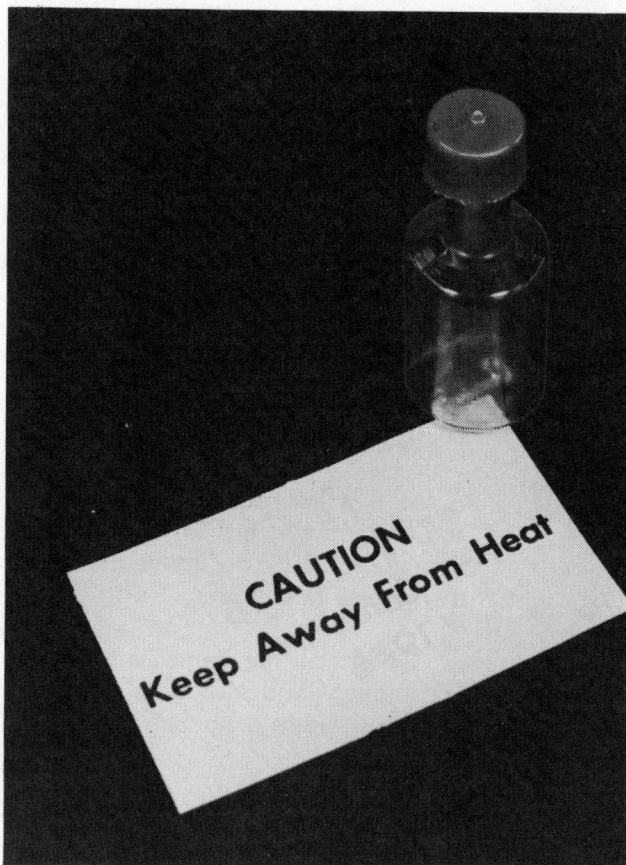

Fig. 30-3   Product to be impulse sealed.

Fig. 30-4 Foot pedal assembly connected to impulse sealer.

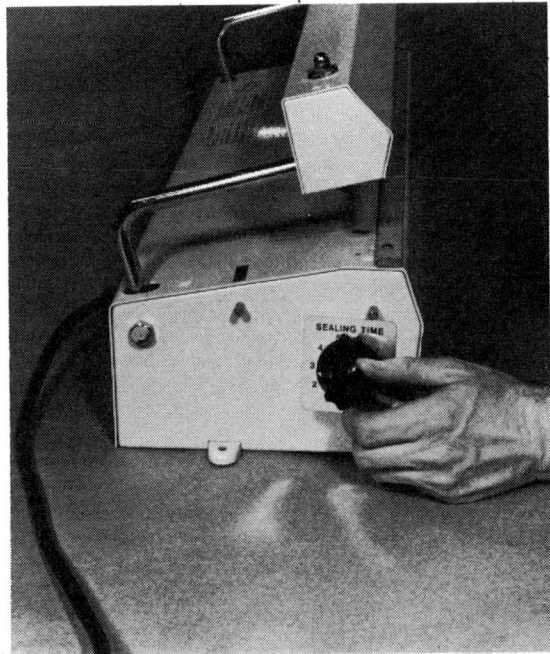

Fig. 30-5 Adjusting timer.

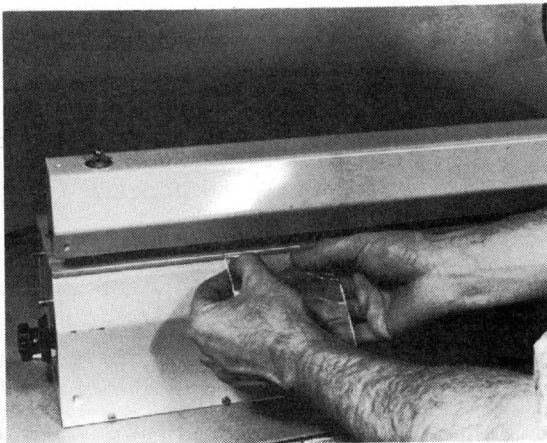

Fig. 30-6 Sealing edges of package.

Fig. 30-7 Sealing open end.

Impulse Sealing

14. Seal the open package end as illustrated in Step 11. See Fig. 30-7.

15. Look at the completed, impulse-sealed package. Pull the sealed joint open. See Fig. 30-8. A well sealed joint will be free of wrinkles. It will also not separate, but will tear next to the seal. If the seal seems to pull apart **too** easily, increase the sealing time. **Reduce** the sealing time and **increase** the cooling time if the seal is melted and torn when removed from the sealer.

16. Make a few sealing tests before packaging a finished product. This will help you find the proper sealing and cooling time for each type of plastic.

CONCLUSION

The impulse sealing process just described is hand-fed and semiautomatic. It is used in school laboratories to package custom made products. It is also used to package products mass produced by industry. This is done by altering the machine, plastic feeding unit, and the plastic cutting unit. Fig. 30-9 shows a number of products with impulse sealed packages.

REVIEW

1. Name the plastic sealed in this activity.

2. Explain each step in the impulse sealing process.

3. Describe how the impulse sealing machine works.

4. Name two types of plastic that may be joined by impulse sealing.

5. Name three products whose packages were impulse sealed.

6. Show what a good seal looks like by tearing one apart.

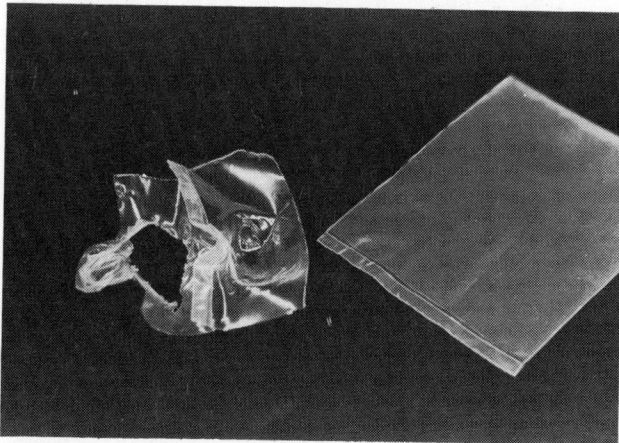

Fig. 30-8   Sealed joint pulled open.

Fig. 30-9   Examples of impulse sealed products.

7. Describe how to adjust equipment when a melted and wrinkled seal is made.

8. Describe how to adjust equipment when a joint did not seal.

9. Describe how to adjust equipment when plastic sticks to the sealing jaw.

SELECTED BIBLIOGRAPHY

Agranoff, Joan, ed. *Modern Plastics Encyclopedia.* New York: McGraw-Hill Book Company, 1976-77.

Baird, Ronald J. *Industrial Plastics.* South Holland, Illinois: The Goodheart-Willcox Company, Inc., 1971.

Briston, J. H. *Plastics Films.* New York: John Wiley & Sons, 1974.

Milby, Robert V. *Plastics Technology.* New York: McGraw-Hill Book Company, 1973.

Park, W. R. R., ed. *Plastics Film Technology.* Huntington, New York: Robert E. Krieger Publishing Company, 1973.

Patton, William J. *Plastics Technology: Theory, Design, and Manufacture.* Reston, Virginia: Reston Publishing Company, Inc., 1976.

EQUIPMENT AND MATERIAL SUPPLIERS

1. Delvie's Plastics, Inc., 2320 South West Temple, P.O. Box 1415, Salt Lake City, Utah 84110.

2. Graves-Humphreys, Inc., P.O. Box 13407, 1948 Franklin Road, Roanoke, Virginia 24033.

Impulse Sealing

3. McKilligan Industrial Supply Corporation, 494 Chenango Street, Binghamton, New York 13901.

4. Paxton/Patterson, 5719 West 65th Street, Chicago, Illinois 60638.

The equipment and materials used in impulse sealing can be purchased from any of the suppliers listed above.

Assignment 31
Hot Wire Sealing

OBJECTIVES

To prepare hot wire sealing equipment, plastic, and products.

To make a hot wire sealed package.

INTRODUCTION

Hot wire sealing is a heat sealing process. It is used by the packaging industry to seal products in plastic. Hot wire sealing machines seal and cut the plastic at the same time. Hot wire sealing plastics include nylon, vinyl, polyethylene, polypropylene, and fluorocarbons. Hot wire sealed packages are used to protect records, toys, games, books, electrical devices, food, auto parts, and tools. This process is also used to make bags around products to be "heat shrink" packaged.

The hot wire sealing process works as follows. The edges of two pieces of plastic are held together. They are placed between a movable and a stationary set of jaws in a sealing machine. The upper (movable) jaw is brought down on top of the two pieces of plastic. The pieces are then tightly pressed together against the stationary sealing jaw.

An electrical element is embedded in the stationary jaw. When the two jaws make contact, a switch for the element is turned on. Current flows in the wire element and makes heat for as long as the jaws make contact. This heat bonds the two pieces of plastic together. The upper jaw is then opened a little. This turns off the current and lets the plastic cool. Then, the upper jaw is completely opened and the sealed plastic may be removed.

Hot Wire Sealing

The plastic is often sealed and cut at the same time. This makes a sealed package and the first sealed end of the next package.

SAFETY

The following precautions should be taken when hot wire sealing:

1. Wear safety glasses.

2. Learn the safe operation of the hot wire sealer.

3. Do not pinch your hands and fingers between the movable upper jaw and the stationary sealing jaw.

4. Use the proper sealing time, sealing pressure, and cooling time for the plastic being sealed.

5. Learn how to properly assemble, adjust, and use the hot wire sealing attachments.

EQUIPMENT AND MATERIALS

The equipment and materials needed for hot wire sealing are:

1. Safety glasses.
2. Hot wire sealer.
3. Nylon, vinyl, polyethylene, polypropylene, or fluorocarbon film sealing plastic.
4. Sealing plastic feed mechanism.
5. Product to be packaged.
6. Scissors.

HOT WIRE SEALING LAB PROCEDURE

A small, school laboratory hot wire sealer is shown in Fig. 31-1. Your school shop may have a similar one. Be sure to understand how your sealer works before using it.

Section X: Packaging

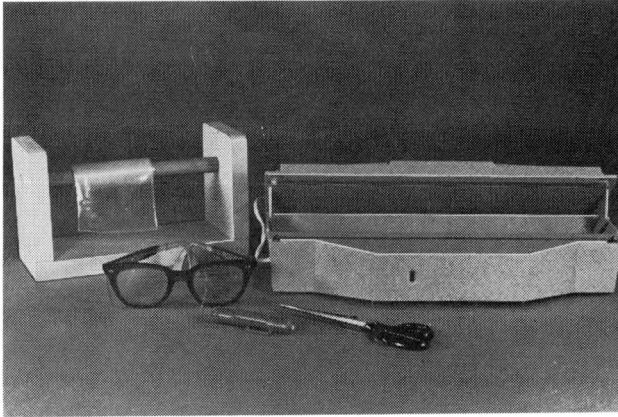

Fig. 31-1 Hot wire sealing equipment and materials.

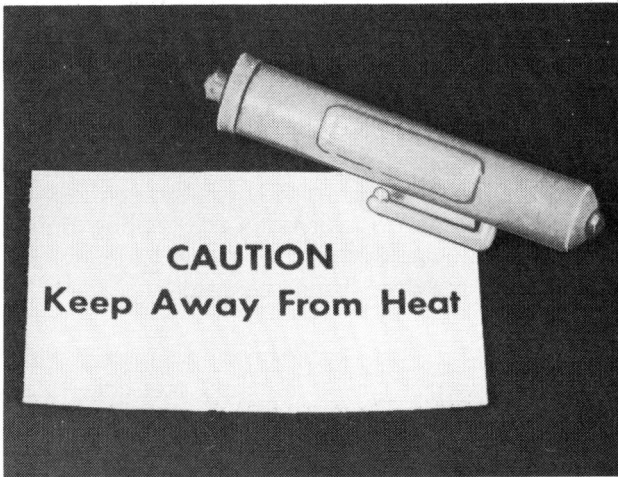

CAUTION
Keep Away From Heat

Fig. 31-2 Product to be hot wire sealed.

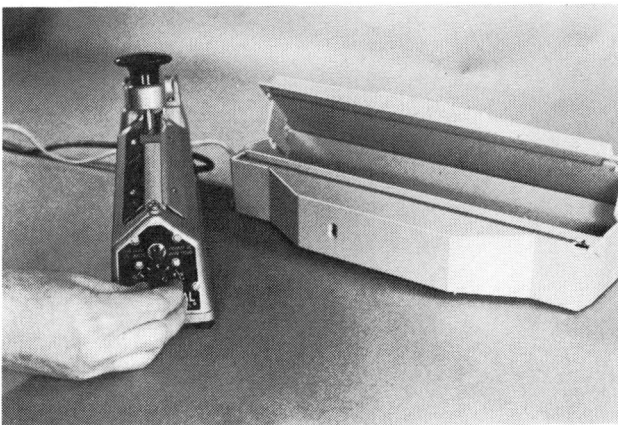

Fig. 31-3 Adjusting timer.

1. Obtain all the items listed under "Equipment and Materials" needed to hot wire seal a package. See Fig. 31-1.

2. Obtain the products to be hot wire seal packaged. See Fig. 31-2.

3. Plug in the electrical cord of the sealer. Some hot wire sealers have a timer that must be adjusted. See Fig. 31-3. The timer adjustment is demonstrated in Assignment 30 (Impulse Sealing). Follow your instructor's recommendations for the proper sealing time. Generally, the thicker the plastic, the longer the sealing time.

4. Hold the partial plastic package between the upper jaw and the bottom or hot wire jaw.

5. Press the upper jaw snugly against the bottom jaw. See Fig. 31-4. Ask your instructor how long to hold the jaws together. As long as the jaws are together, the wire heats and stays hot. An impulse sealer wire will stay hot only for the amount of time set.

6. Release enough pressure on the upper jaw to turn off the electric current. This will let the wire and the joint cool.

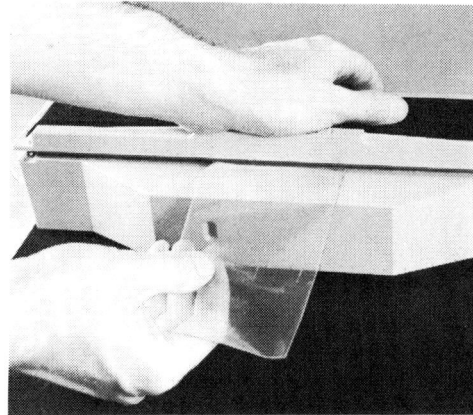

Fig. 31-4 Pressing upper jaw against lower jaw.

Hot Wire Sealing

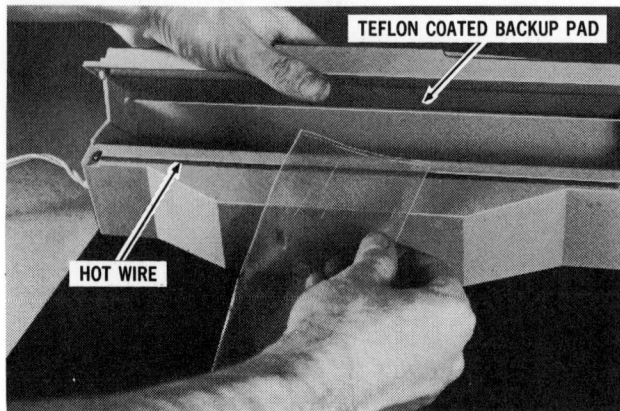

Fig. 31-5 Opening upper jaw to remove plastic.

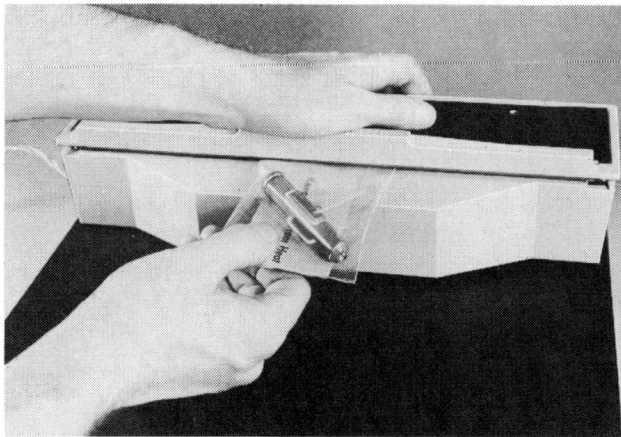

Fig. 31-6 Sealing open package end.

Fig. 31-7 Hot wire sealed package.

7. Keep a light amount of pressure on the jaws for the same amount of time as was used in Step 5. This lets the joint cool. If the sealer has a timer, follow the procedure demonstrated in Assignment 30.

8. Open the upper jaw to remove the plastic. See Fig. 31-5. If the plastic sticks to the jaws, wipe the jaws clean and rub a silicone paste release agent on each jaw.

9. Place the product(s) in the partial package.

10. Seal the open package end. See Steps 5, 6, 7, and 8. See Fig. 31.6.

11. Fig. 31-7 shows the finished, hot-wire sealed package. Pull the sealed joint open. A well sealed joint will be free of wrinkles. It will also not separate, but will tear next to the seal. If the seal seems to pull apart too easily, increase the sealing time. Reduce the sealing time and increase the cooling time if the seal is melted and torn when removed from the sealer.

12. Make a few sealing tests before packaging a finished product. Note that the thin round wire cools very fast. This helps the seal cool fast and speeds up the sealing process.

CONCLUSION

A number of products in hot wire sealed packages are shown in Fig. 31-8. Use the process just described to make these packages. This process is a hand-fed setup. It is used in school laboratories to package custom made projects. It is also used by industry to package mass produced products. This is done by changing the sealing machine plastic, feeding unit, and the plastic cutting unit.

Section X: Packaging

Two heat sealing units are shown in Fig. 31-9. The unit on the left is a flat element impulse sealer. The unit on the right is a round element (wire) hot wire sealer. The flat element unit makes a wide seal. The wire element unit makes a narrow seal.

Often, a straight edge sealer is replaced with an "L" type hot wire sealer. This allows two sealing units to be placed at right angles to one another. This also allows two sides of a plastic package to be sealed at one time. For this process, a continuous piece of plastic, pre-folded in half, is used. This plastic is used by the roll. Any product packaged before this has left the end of the plastic sealed. The operator places the product in the half formed package. (The side at the end of the plastic roll is sealed and the other side is the fold.) The operator brings the "L" sealer down on the plastic to complete the package sealing and cutting. This hot wire "L" sealer speeds up the packaging process. A product is bagged, sealed, and cut in one step.

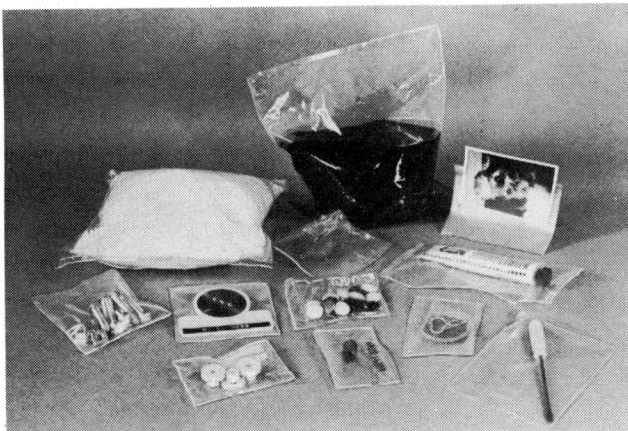

Fig. 31-8 Examples of hot wire sealed packages.

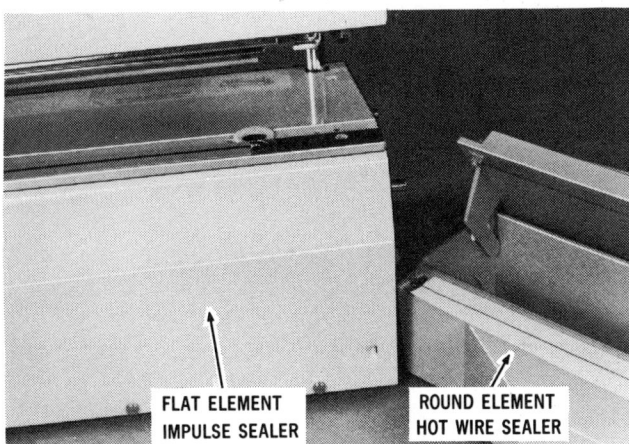

FLAT ELEMENT
IMPULSE SEALER

ROUND ELEMENT
HOT WIRE SEALER

Fig. 31-9 Two types of heat sealing units.

REVIEW

1. Name the plastic sealed in this activity.

2. Describe each step in the hot wire sealing process.

3. Describe how the hot wire sealing machine works.

4. Name two types of plastic that may be joined by hot wire sealing.

5. Name three products whose packages were hot wire sealed.

6. Show what a good seal looks like by tearing one apart.

Hot Wire Sealing

7. Describe how to adjust the hot wire sealer when a melted and wrinkled seal is made.

8. Describe how to adjust equipment when a joint does not seal.

9. Describe how to adjust equipment when plastic sticks to the sealing jaw.

10. Explain the use of an "L" type hot wire sealer.

SELECTED BIBLIOGRAPHY

Briston, J.H., *Plastics Films.* John Wiley and Sons, New York, 1974.

Park W.R.R., ed. *Plastics Film Technology*, Robert E. Krieger Publishing Company, Huntington, New York, 1973.

Patton, William J. *Plastics Technology: Theory, Design, and Manufacture.* Reston, Virginia: Reston Publishing Company, Inc., 1976.

EQUIPMENT AND MATERIAL SUPPLIERS

1. Allied Automation Inc., 9144 King Arthur Drive, Dallas, Texas 75247.

2. Beseler Corporation, 8 Fernwood Road, Florham Park, New Jersey 07932.

3. Clamco Corporation, 11350 Brookpark Road, Cleveland, Ohio 44130.

4. Laminaire Corporation, 1072 Randolph Avenue, Rahway, New Jersey 07065.

5. Sears Roebuck & Company, Sears Tower, Chicago, Illinois 60684.

6. Weldotron Corporation, 1532 South Washington Avenue, Piscataway, New Jersey 08854.

Commercial hot wire sealing equipment (round and flat wire element types) can be purchased from suppliers 1, 2, 3, 4, and 6. Hot wire impulse sealing equipment (flat wire element type) and plastic can be purchased from any of the suppliers listed in Assignment 30 (Impulse Sealing). Any non-commercial and light-duty hot wire sealing equipment and supplies can be purchased from Sears Roebuck & Company (see above).

Hot Wire Sealing

Assignment 32
Heat Shrink Tunnel Sealing

OBJECTIVES

To prepare a product and heat shrinking equipment.

To heat shrink package a product.

INTRODUCTION

Heat shrink tunnel sealing is used to make a package fit more snugly around a product. The plastic used is a special heat shrink material. It is designed to shrink 10 to 50% of its original size.

Polyolefins, fluoroplastics, and polyvinyls are the plastics commonly used. These plastics are stretched during their initial processing. Then, when they are again heated, they shrink (return) to their original size.

Shown in Fig. 32-1 is a small, industrial heat shrink tunnel and heat sealer. Products (computer cards) are wrapped with plastic fed from a roll. The plastic is impulse sealed around the product. The package is then placed on a moving belt. This belt moves the package through the heat tunnel. The heat causes the plastic to shrink tightly around the product.

The heat shrink tunnel technique is used to package cartons, printed material, food, toys, records, games, tools, automobile parts, and other products. This technique protects the product from dirt, bacteria, freezer burn, and theft. It also results in less storage space.

Fig. 32-1 Industrial heat shrink tunnel and heat sealer. (Courtesy Weldotron Corporation)

Section X: Packaging

SAFETY

The following precautions should be taken when heat shrink tunnel packaging:

1. Wear safety glasses.

2. Work on a heat-resistant surface.

3. Wear heat-resistant gloves.

4. Learn how to safely operate and adjust the heat shrink tunnel.

5. Keep your hands and fingers out of the heat shrink tunnel.

6. Learn how to safely operate the tacking tool.

7. Place the hot tacking tool only on a heat-resistant surface.

8. Learn how to safely operate the hair dryer.

EQUIPMENT AND MATERIALS

The equipment and materials needed for heat shrink tunnel packaging are:

1. Safety glasses.
2. Heat–resistant gloves.
3. Heat–resistant surface.
4. Heat sealer.
5. Heat shrink tunnel.
6. Hair dryer.
7. Polyolefin, fluorocarbon, or polyvinyl chloride heat shrinking plastic.
8. Products to be packaged.
9. Tacking tool.
10. Scissors.
11. Rule.

HEAT SHRINK TUNNEL SEALING LAB PROCEDURE

Shown in Fig. 32-2 is a common school laboratory heat shrink

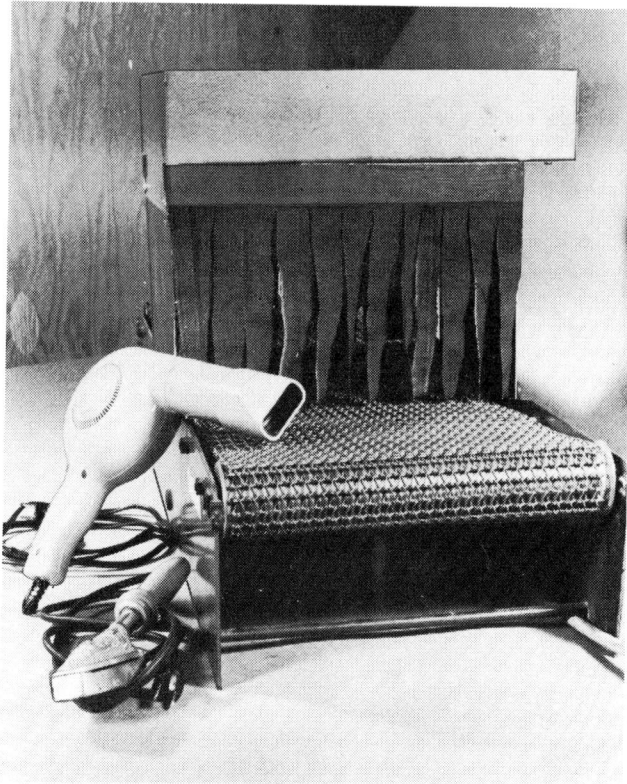

Fig. 32-2   School laboratory heat shrink packaging setup.

Heat Shrink Tunnel Sealing

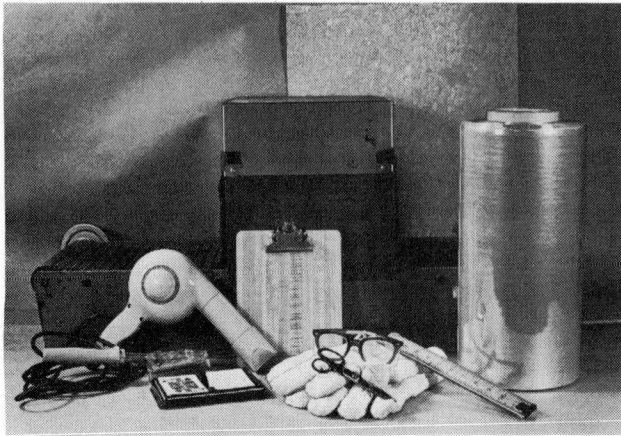

Fig. 32-3 Heat shrink packaging equipment and materials.

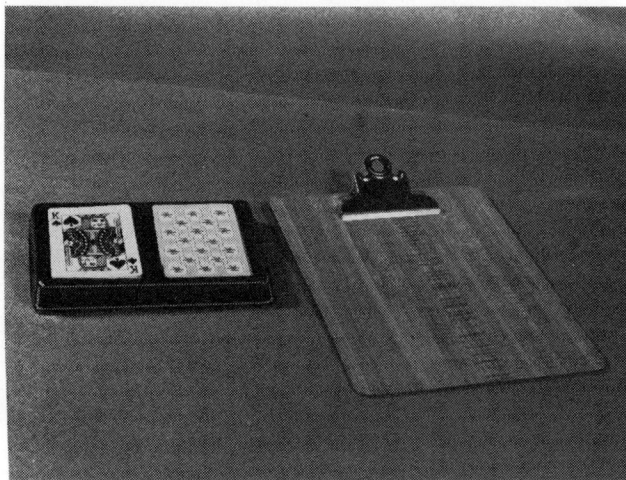

Fig. 32-4 Products to be heat shrink packaged.

Fig. 32-5 Wrapping the product.

packaging setup. Your school and many small businesses may have similar equipment. Be sure you understand all the parts of the equipment before beginning this assignment.

1. Obtain all the items listed under "Equipment and Materials" needed to heat shrink a package. See Fig. 32-3.

2. Obtain the products to be heat shrink packaged. See Fig. 32-4.

3. Cut a piece of heat shrink plastic large enough to wrap the product. Ask your instructor what size to cut the plastic.

4. Wrap the plastic around the product. See Fig. 32-5. Wrap the product so that the plastic edges meet on the **bottom** of the product. Place the product upside-down for wrapping.

5. Make the plastic edges join on the bottom of the product. See Fig. 32-6.

6. Hold a warm tacking tool snugly against the package seam. See Fig. 32-7. Apply heat to the seam for only a second or two. Heat sealing time and temperature is different for each plastic and plastic thickness. Ask your instructor to tell you what sealing time and temperature to use. The heat and pressure of the tacking tool will cause the plastic to melt and bond together.

7. Direct warm air from a hair dryer or a heat gun all around the plastic. The plastic will shrink tightly around the product. **Do not melt the plastic by using too much heat.** Be careful not to burn yourself with the hot air. (Use your instructor's recommendations for the hair dryer heat settings. The heat range may vary from 250 to 350°F. This depends on the type and thickness of the plastic.)

Fig. 32-6 Joining the edges.

Fig. 32-7 Holding warm tacking tool against plastic seam.

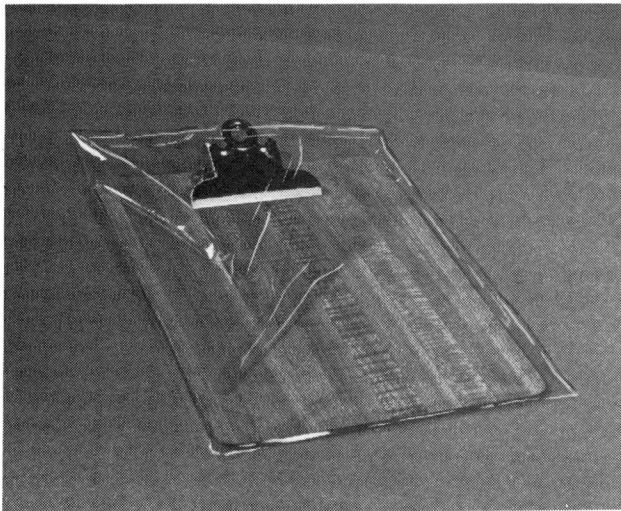

Fig. 32-8 Heat sealing loose plastic bag around clip board.

8. Heat seal a loose plastic bag around a product. See Fig. 32-8. Follow the procedures demonstrated in Assignments 16, 30, and 31. Ask your instructor what plastic type and bag size to use. (Heat shrink plastic is sold in different thicknesses and "percent of shrink" ranges.)

9. Turn on the heat shrink tunnel heater switch. See Fig. 32-9.

10. Turn on the belt and adjust its speed. See Fig. 32-10. Ask your instructor what speed setting to use. The belt speed depends on how much shrinkage you want to take place. It also depends on how many products you want to heat shrink.

Fig. 32-9 Turning on heat shrink tunnel heater.

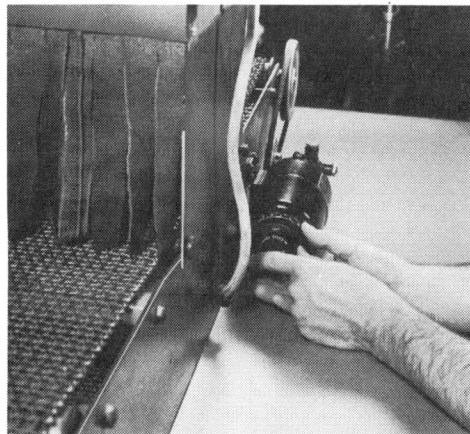

Fig. 32-10 Turning on belt and adjusting its speed.

Heat Shrink Tunnel Sealing

(Sometimes, a small hole is punched in the bottom of the package. This lets air escape from the package while the plastic is shrinking.)

11. Place the package on the moving belt.

12. Remove the finished heat shrink tunnel package from the belt.

CONCLUSION

Fig. 32-11 shows two heat shrink packaged products. Each product was heat shrink packaged with a different process. Shown in Fig. 32-12 are many products that were also heat shrink packaged. The procedure just described can be used to package these products.

The heat shrink process described here is slow. By altering heat shrink equipment, industry uses these techniques to mass package products. Often, industrial heat shrink tunnels can take 45 to 150 packages per minute.

Heat shrink tunnels can be custom built. Sheet metal is used for the tunnel. A hot air gun is the heat source. A metal mesh belt can be made to run the entire length of the tunnel. The belt can be either hand or motor operated.

Products that protect electrical splices, electrical components, or symmetrical machine parts are heat shrink packaged. The electrical splices, electrical components, or machine parts are placed in a plastic heat shrinking tube. The tube is heated in the heat shrink tunnel. Hot air causes the tube to shrink tightly around the part. Heat shrink tubing is often called "spaghetti" or "shrink-fit covers."

REVIEW

1. Name the heat shrinking plastic used in this unit.

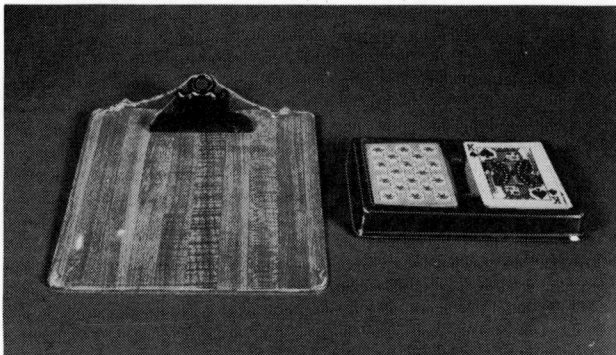

Fig. 32-11 Two heat shrink packaged products.

Fig. 32-12 Other heat shrink packaged products.

Section X: Packaging

2. Describe each step in the heat shrink tunnel packaging process.

3. Show how the heat shrink tunnel works.

4. Explain why heat shrink film shrinks.

5. Name three products with heat shrink packages.

6. Explain where a *tacker* is used.

7. Describe how to adjust the heat shrink tunnel.

8. List three plastics used for heat shrink film tubing.

9. Describe a use for heat shrink tubing.

10. List three advantages for heat shrink product packaging.

SELECTED BIBLIOGRAPHY

Briston, J.H., *Plastic Films.* John Wiley and Sons, New York, 1974.

Park, W.R.R., ed. *Plastics Film Technology,* Robert E. Krieger Publishing Company, Huntington, New York, 1973.

Patton, William J. *Plastics Technology: Theory, Design, and Manufacture.* Reston, Virginia: Reston Publishing Company Inc., 1976.

Richardson, Terry A. *Modern Industrial Plastics.* Indianapolis: Howard W. Sams & Company, Inc., 1974.

EQUIPMENT AND MATERIAL SUPPLIERS

1. Ain Plastics Inc., 160 MacQuesten Parkway So., Mt. Vernon, New York 10550.

Heat Shrink Tunnel Sealing

2. Allied Automation, Inc., 9144 King Arthur Drive, Dallas, Texas 75247.

3. Allied Chemical Corporation, Specialty Chemical Division, P.O. Box 1057R, Morristown, New Jersey 07960.

4. Beseler Corporation, 8 Fernwood Road, Florham Park, New Jersey 07932.

5. Cadillac Plastic and Chemical Co., P.O. Box 810, Detroit, Michigan 48232.

6. Clamco Corporation, 11350 Brookpark Road, Cleveland, Ohio 44130.

7. Errich International Corporation, Errich Packaging Machine Division, 721 Broadway, New York 10003.

8. Plastic Manufacturers Inc., 4041 Ridge Ave., Bldg. 31, Philadelphia, Pennsylvania 19129.

9. Weldotron Corporation, 1532 So. Washington Ave., Piscataway, New Jersey 08854.

Heat shrink tunnels are available from suppliers 2, 4, 6, 7, and 9. Heat shrinking plastic can be purchased from suppliers 1, 3, 5, and 8.

Section XI
Ecology

Assignment 33
Recycling

OBJECTIVES

To collect, identify, and separate one type of thermoset plastic found in household, industrial, and school trash.

To collect, identify, and separate three types of thermoplastics found in household, industrial, and school trash.

To regrind and store the collected plastic.

To make products with a filler of recycled plastic.

INTRODUCTION

The amount of plastics in solid waste must be controlled. Recycling scrap plastic is one solution. Recycling requires that different scrap plastics be identified, separated and cleaned. However, there is no cheap way to do it.

A typical solid waste disposal plant is shown in Fig. 33-1. Here, plastic products are burned with other refuse such as packaging materials and food scraps. In thermoplastic processing plants, recycling is a common practice. Thermoplastic scrap is separated, cleaned and granulated. It is then mixed with new plastic and molded into various products.

Many industrial, educational, and governmental agencies are researching methods for recycling plastic scrap. Some of the recycling experiments involve producing power and fuel. Others involve making foamed products, shipping pallets, chemicals, drain pipes, and lawn sprinklers from plastic waste.

Fig. 33-1 Typical solid waste disposal plant.

Recovering urethane foam, polyester, PVC resin and ABS from scrap plastic is still another experiment. A few recycling methods are described in this assignment.

SAFETY

The following precautions should be taken when recycling plastic:

1. Wear safety glasses.

2. Wear gloves while collecting and separating trash.

3. Learn how to safely operate the granulator.

4. Do not place your hands and fingers into the feed area of the granulator.

5. Learn how to safely operate all plastic processing equipment used for recycling plastic into finished products (Examples: extruder and injection machines).

6. Avoid contacting or breathing the harmful chemicals produced by trash degrading and plastic recycling.

7. Do all plastic recycling operations in a well-ventilated area.

8. Unplug the granulator before opening it.

9. Feed plastic slowly into the granulator to avoid jamming it.

EQUIPMENT AND MATERIALS

The equipment and materials needed for plastic recycling are:

1. Safety glasses.
2. Gloves.
3. Trash containing scrap plastic.

Recycling

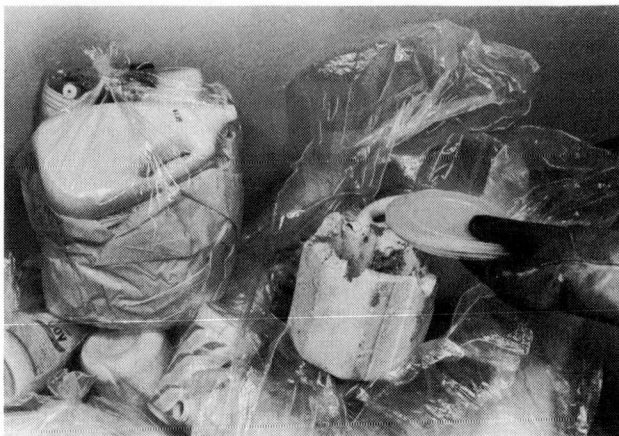

Fig. 33-2   Separating plastic from nonplastic trash.

Fig. 33-3   Removing ferrous metals.

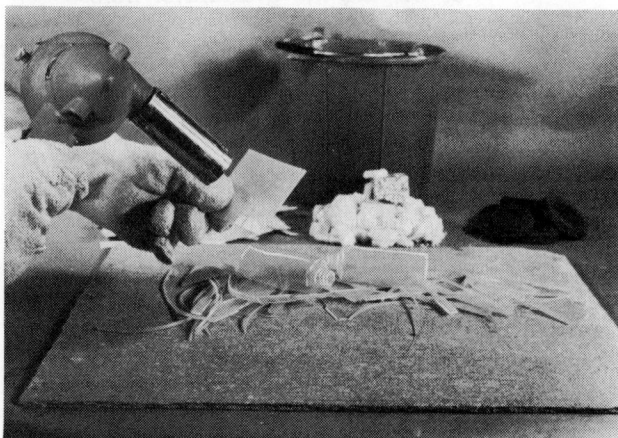

Fig. 33-4   Drying plastic with heat gun.

4. Granulator.
5. Heat gun.
6. Magnet.
7. Scissors.
8. Metal plate.
9. Hammer.
10. Bucket of water.
11. Cloth sack.

RECYCLING LAB PROCEDURE

Several methods of separating, identifying, and recycling scrap plastics found in solid wastes are described in this section.

1. Collect household, industrial, and school trash containing plastic.

2. Separate the plastic from the nonplastic trash. See Fig. 33-2. Wear gloves and work in a well-ventilated area.

3. Identify the different plastics in the trash.

4. Separate each type of plastic into its own pile. Polyethylene is in the front, styrene on the left, foamed styrene in the center, and phenolic on the right. Some plastics will be hard to identify. Use the methods demonstrated in Assignments 1 and 2 of this manual. Limit your sorted trash to one thermoset and three thermoplastic types. Plastics easily collected from trash are phenolic (thermoset), polyethylene, polystyrene, and foamed polystyrene (thermoplastics).

5. Move a magnet slowly through the piles of plastic. See Fig. 33-3. This removes ferrous (iron containing) metals from the plastic.

6. Clean each pile of plastic with water.

7. Dry the plastic with a heat gun. Fig. 33-4.

Fig. 33-5 Placing polyethylene plastic in granulator.

Fig. 33-6 Emptying granulator drawer.

8. Turn on the granulator.

9. Place small plastic pieces from one pile (polyethylene) slowly into the granulator. See Fig. 33-5. Large pieces must be cut into small pieces before placing them into the granulator. Sheet plastic (thermoforming scrap) must be cut into small squares before placing them in the granulator. Feed the granulator only small amounts of scrap at a time. Let the granulator grind all the scraps before feeding it more. (This prevents jamming the granulator.) *Do not place your hands or fingers in the granulator feed opening.*

10. Grind all of the polyethylene scrap until the granulator drawer is full.

11. Turn off the granulator and unplug it.

12. Empty the granulator drawer on a bench. See Fig. 33-6.

13. Pass the magnet through the ground plastic. This removes any ferrous metal that would damage polyethylene processing equipment. The polyethylene is now ready to recycle.

14. Clean and granulate the polystyrene scrap. (Use the procedure just demonstrated for polyethylene.) See Fig. 33-7. Always clean the inside granulator parts before grinding a different plastic. This prevents one plastic from being mixed with another type. Follow your instructor's directions for cleaning the granulator. Make sure you unplug the granulator before opening it.

15. Place the phenolic scrap in a sack or wrap it in a cloth.

16. Lay the sack of scrap on a metal plate.

Recycling

Fig. 33-7 Placing polystyrene plastic in granulator.

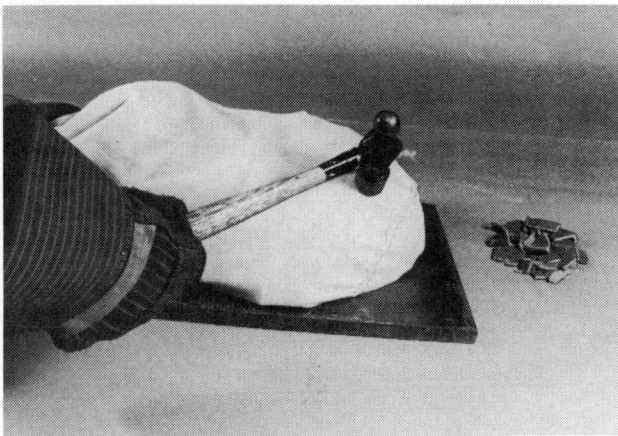

Fig. 33-8 Hammering plastic into small pieces.

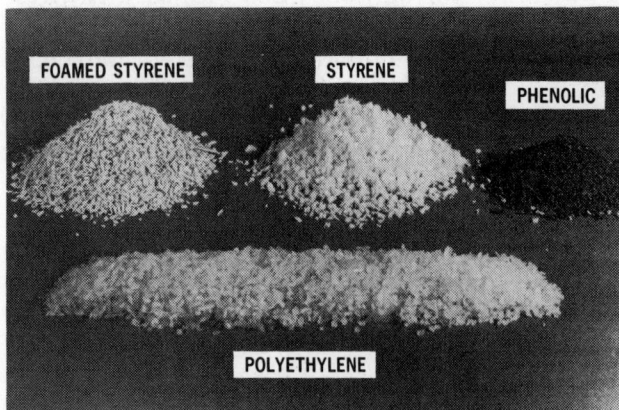

Fig. 33-9 Plastics ready for recycling.

17. Reduce the size of the plastic by hammering it into small pieces. See Fig. 33-8. *Be sure to wear safety glasses.*

18. Clean and granulate the phenolic scrap as was demonstrated in Steps 6 through 13. The expanded polystyrene can also be granulated. It may also be cut into small pieces with scissors.

19. Note the four piles of plastic shown in Fig. 33-9. Each pile contains a different plastic ready for recycling.

CONCLUSION

Polyethylene, polystyrene, phenolic, and foamed polystyrene products are shown in Fig. 33-10. Each product contains as a filler one of the scrap plastics shown in Fig. 33-9.

Recycled polyethylene. About 7% to 50% recycled polyethylene scrap can be mixed with new polyethylene and extruded into new products. An example is drainage pipe. This mixture of recycled and new polyethylene can also be pulverized or crushed. It can then be used to make rotationally molded products. The same mixture of recycled and new polyethylene can be mixed with colorants. It is then injected to form bulletin board tacks and other products.

One hundred percent recycled polyethylene scrap can be used in thermofusion projects. Recycled polyethylene can be compressed under heat and pressure in a laminating press. This forms small sheets of plastic for thermoforming. Recycled polyethylene can also be mixed with sawdust and compressed into a solid fuel wafer. The wafer is very combustible and produces a great amount of heat.

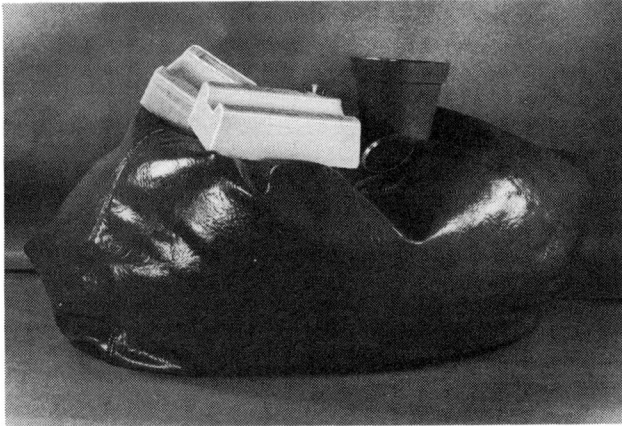

Fig. 33-10 Products made from recycled plastics.

SLUDGE

POLYETHYLENE

MULCH

SAND AND LOAM

Fig. 33-11 Mulch mixture containing plastic.

Fig. 33-12 Vegetables transplanted through degradable mulch plastic. (Courtesy Princeton Chemical Research, Inc.)

A plant mulch can be formed by mixing one part granulated polyethylene with one part sandy loam and ½ part sludge. See Fig. 33-11. Sludge can be obtained from some sewage treatment plants. The polyethylene in the mulch prevents the soil from packing. This allows air to circulate through the mixture and quickens the decay of organic matter. Other mulch mixtures can be experimented with. Shown in Fig. 33-12 are vegetables transplanted through a *degradable* mulch plastic. The plastic is programmed to degrade when exposed to sunlight.

Fig. 33-13 shows kale growing under controlled laboratory conditions. It is growing in a soil-less base of various layers of recycled polyethylene. Liquid foods are circulated through the plastic and the plant roots. This plant growth method is called "hydroponics." Polyethylene makes a good base for the roots because it does not degrade.

Fig. 33-13 Polyethylene plastic used as plant base.

Recycling

Recycled polystyrene. About 7% to 50% recycled polystyrene scrap can be mixed with new polystyrene and colorants. It can be injected in the same way as the polyethylene mixture to form new products.

Recycled polystyrene can also be extruded and compressed as was described for recycled polyethylene. One hundred percent recycled polystyrene can also be used in thermofusion projects.

Recycled foam polystyrene can be used as a fill material for bean bag furniture. A mixture of about 40% recycled polystyrene foam and 60% new urethane foam can be used to make car top carriers.

Recycled phenolic. Recycled phenolic scrap may be used as a partial sand replacement in concrete. Fig. 33-14 shows the making of concrete residential walk blocks. About 25% of the sand in the concrete was replaced with recycled phenolic plastic. Recycled polyethylene or polystyrene can also be used. Concrete made from recycled plastic can be used as curbing or ornamental concrete.

Recycled phenolic scrap can be used to make compression molded products such as coasters. About 10% or less recycled phenolic scrap is mixed as filler with new phenolic for this purpose.

REVIEW

1. Describe how plastic is recycled in thermoplastic molding plants.

2. Explain what a granulator does.

3. Describe how to separate and identify general purpose phenolic.

4. Explain the reason for cleaning, drying, and sifting plastic trash with a magnet.

Fig. 33-14   Concrete walk blocks made with recycled plastic.

5. Describe one recycling use for general purpose phenolic.

6. Give six recycling uses for polyethylene.

7. Give six recycling uses for polystyrene.

8. Describe two recycling uses for foamed polystyrene.

9. Explain the reason for using specific recycled plastics in mulch.

10. Explain the reason for using specific recycled plastics in hydroponics.

11. Explain the reason for using recycled plastic in concrete.

SELECTED BIBLIOGRAPHY

Richardson, Terry A. *Modern Industrial Plastics*. Indianapolis: Howard W. Sams & Company, Inc., 1974.

Rosato, Dominick V., ed. *Plastics Industry Safety* Handbook. Boston: Cahners Books, 1973.

Articles:
"Plastics," Fact Sheet Bulletin No. NCF-14-01. Washington, D.C: National Center for Resource Recovery, Inc. October 1973.

"Solid Waste Disposal," Fact Sheet Bulletin No. NCF-11-02. Washington, D.C.: National Center for Resource Recovery, Inc., March, 1973.

"Voluntary Separation," Fact Sheet Bulletin No. NCF-19-01. Washington, D.C.: National Center for Resource Recovery, Inc., August, 1973.

EQUIPMENT AND MATERIAL SUPPLIERS

1. Acme Plastic Machinery Corporation, 500 Saw Mill River

Road, Yonkers, New York 10701.

2. Brodhead-Garrett, 4560 East 71st Street, Cleveland, Ohio 44105.

3. California Pellet Mill Company, 1114E Wabash Ave., Crawfordsville, Indiana 47933.

4. Conair, Inc., Conair Building, Franklin, Pennsylvania 16323.

5. Delvies Plastics, Inc., 2320 South West Temple P.O. Box 1415, Salt Lake City, Utah 84110.

6. McKilligan Industrial Supply Corporation, 494 Chenango Street, Binghamton, New York 13901.

7. Rietz Division, Bepex Corporation, 150 Todd Road, Santa Rosa, California 95401.

The plastic for this assignment can be obtained from household, industrial, and school trash. Granulators are available from any of the suppliers listed above. The suppliers of plastic processing equipment are listed in other assignments. For example, the suppliers of injection machines are listed in Assignment 3 (Injection Molding). See the appropriate assignment.

Assignment 34
Disposing

OBJECTIVES

To identify and explain the differences among area, trench, and ramp type sanitary landfills.

To identify and explain the differences among water-wall, pyrolysis, total incineration, and fluid-bed types of incineration.

To describe how plastic solid waste is disposed of in landfill and incineration methods.

INTRODUCTION

The volume of plastic in our solid waste is increasing. Plastics make up about two to three percent of the nation's collected solid waste. Most plastics in solid waste come from discarded items. These items include packaging materials, industrial wastes, home products, furniture, and toys.

Plastics are disposed of in many ways. Some of the disposing methods are biodegradation, composting, open dumping, incineration, and sanitary landfilling. Open dumping has been stopped by many communities. It is an eyesore and a health and fire hazard. It is a poor way to dispose of solid wastes.

Many biodegradable plastics are used today. Biodegradable plastics are those which generally break down under water, soil enzymes, or ultraviolet light. Examples are some pill coatings, some mulch films, some trash bags and some dissolving sutures (surgical stitching). Fig. 34-1 shows a degrading plastic film used as a ground cover for vegetables.

Fig. 34-1 Degrading plastic film. (Courtesy Princeton Chemical Research, Inc.)

Fig. 34-2 Composted mulch mixture.

Fig. 34-3 Older type of incinerator.

Some plastics are used in composting. The plastic does not degrade. This stops the compost mix from compacting. It also allows air to circulate and help break down the organic parts of the compost mix. Only a little plastic is disposed of by composting. Fig. 34-2 shows a composted mulch mixture.

Incineration (controlled burning) is another way to dispose of solid waste. Many incinerators need to be upgraded to meet pollution standards. An older type of incinerator is shown in Fig. 34-3.

Sanitary landfilling is also used to dispose of solid waste. Solid wastes are covered with a layer of dirt. Dirt, broken concrete, bricks, glass, and plastics make good landfill materials. Plastics do not break down or make polluting odors, gases, or liquids. Other organic landfill materials leach out chemicals. These chemicals can contaminate ground water. Fig. 34-4 illustrates a landfill site for a small community.

Many sanitary landfilling and incinerating methods may be used together or separately. Several methods of incineration and sanitary landfilling are described in this assignment.

SAFETY

The following precautions should be taken when visiting solid waste disposal facilities:

1. Wear safety glasses.

2. Wear a hard hat.

3. Wear protective gloves.

4. Wear steel-toed safety shoes.

5. Obey all safety rules posted at the incinerator and landfill sites.

Fig. 34-4 Small sanitary landfill.

Fig. 34-5 Area type landfill.

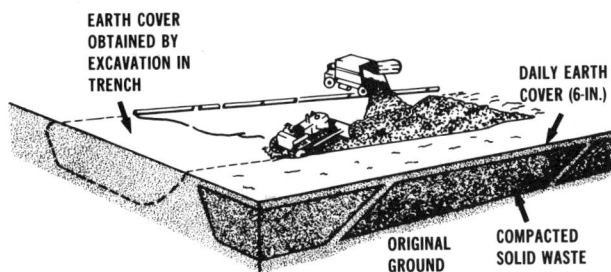
Fig. 34-6 Trench type landfill.

6. Obey all safety rules given by your instructor.

7. Do not wander away from the class.

8. Report any accidents to your instructor or group leader.

9. Do not breathe fumes made by burning or degrading solid waste.

EQUIPMENT AND MATERIALS

The equipment and materials needed for a visit to a solid waste disposal facility are:

1. Safety glasses.
2. Hard hat.
3. Protective gloves.
4. Safety shoes (optional).
5. Pencil.
6. Clipboard and paper.
7. Camera (optional).

FIELD TRIPS

1. Tour an **area type** sanitary landfill. See Fig. 34-5. This type is used on sloping land or low areas. Often quarries, marshes, or canyons are used. Solid waste is dumped next to the landfill area. The waste is spread and compacted with a bulldozer. Dirt is taken from high land and spread over the solid waste. The dirt (6 to 9 inches) is covered and compacted over the solid waste each day or more often. The completely dirt-enclosed solid waste is called a cell. Cells are made side by side and on top of each other until the landfill site is filled.

2. Tour a **trench type** sanitary landfill. See Fig. 34-6. This type is used on flat or slightly sloping land. In this type, many trenches are dug side by side. They are separated by dirt walls. Generally, dirt from a trench is used to fill the landfilled

Disposing

Fig. 34-7 Ramp-type sanitary landfill.

Fig. 34-8 Typical landfill equipment.

Fig. 34-9 Dirt mounds around landfill.

trench. The trenches can be ten feet deep and 50 to 100 feet long. They are at least two tractors wide. The trenches are filled with many dirt enclosed cells of solid waste. The final layer of dirt is two to three feet deep.

3. Tour a **ramp type** sanitary landfill. See Fig. 34-7. This type is used on a steep slope. The solid waste is dumped on the side of the slope. It is spread and compressed with a bulldozer. The dirt is dug from below the dumping site. It is spread over the waste as needed. This method is used by many small communities because only one bulldozer is needed.

4. Look at the machine used to move solid waste in a sanitary landfill. See Fig. 34-8. Compacting stops solid waste from settling, cracking cover soil, and forming gas pockets. Compacting is done by driving heavy equipment over the solid waste. Often this equipment has specially designed wheels to compact the solid waste.

5. Note the dirt mounds around many landfills. See Fig. 34-9. The mounds are used to hide the landfill from the public's view. They also give more depth for extra cover dirt. The mounds are often enclosed by fences to trap litter.

6. Note how some solid waste is milled before being placed in a landfill. See Fig. 34-10. Mills for solid wastes are made in many sizes. Milling solid waste has many advantages. They include the reduced need for landfill cover. They also increase the site life and make easier the separation of solid waste ingredients for resource recovery. Solid wastes like paper, plastic-coated paper, and thin and rigid plastics are often chopped up. They are mixed with water and piped to a dewatering device. They

312

Fig. 34-10 Milling solid waste.

Fig. 34-11 Compacting solid waste.

Fig. 34-12 Former landfill site.

are then partially dried and shipped to a landfill.

7. Note how some solid waste is compacted *(baled)* before being placed into a landfill. See Fig. 34-11. This material is being compressed and loaded into trucks. It will be hauled to landfill sites. High pressure is used to compress either milled or unprocessed solid waste into packages. Baling solid waste has many advantages. These include an increased landfill life, cheaper transportation, and improved landfill operations.

8. Walk over some completed landfill sites. Fig. 34-12 shows one use for landfill sites. Parks and open recreation areas can be built on landfill sites. If buildings are placed on fill sites, they should be lightweight and open. Often explosive methane gases will exhaust from a landfill site. The gas must not be allowed to accumulate in buildings.

9. Tour different incinerator facilities after studying landfill techniques. Fig. 34-13 shows a sketch of one type of water-wall incinerator. The walls of the solid waste burning area are lined with tubes of water. The heat made by the burning waste turns the water to steam. The steam can be used in many ways. It can be used to generate electricity or to heat and cool buildings.

10. Look at the sketch of the incinerator shown in Fig. 34-14. The solid waste is placed in a shredder. Then, the shredded material is roughly separated by air. It is separated into metals and glass which are recycled. The paper, plastic, and yard debris are further separated. The lightest weight materials are shot into the burning chamber of the incinerator. The pressurized gases of the burning

Disposing

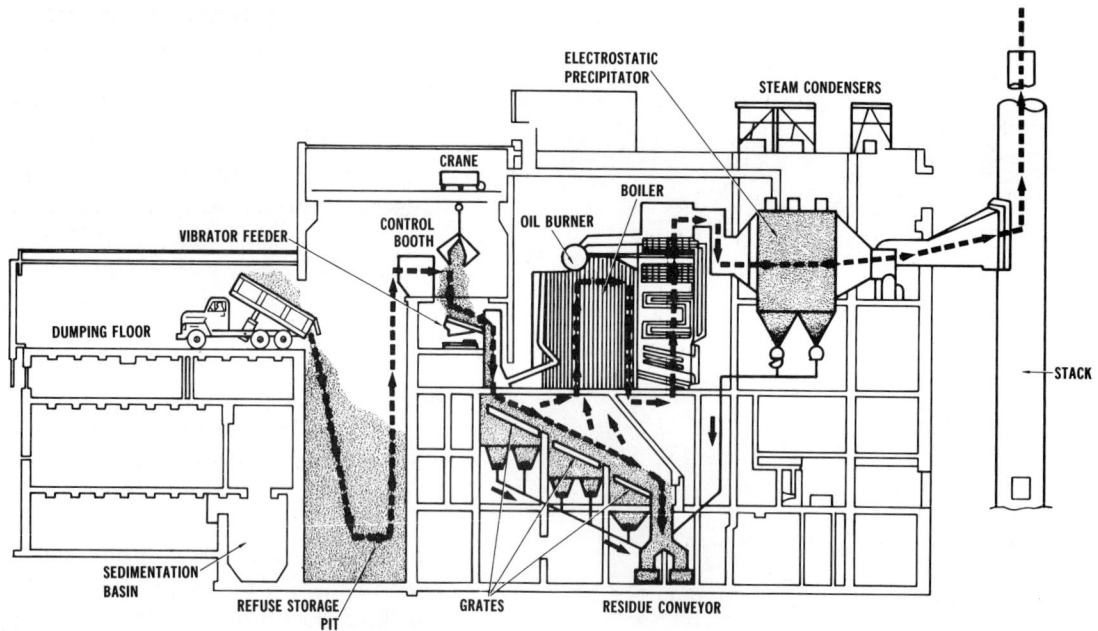

Fig. 34-13 Schematic of water wall incinerator.

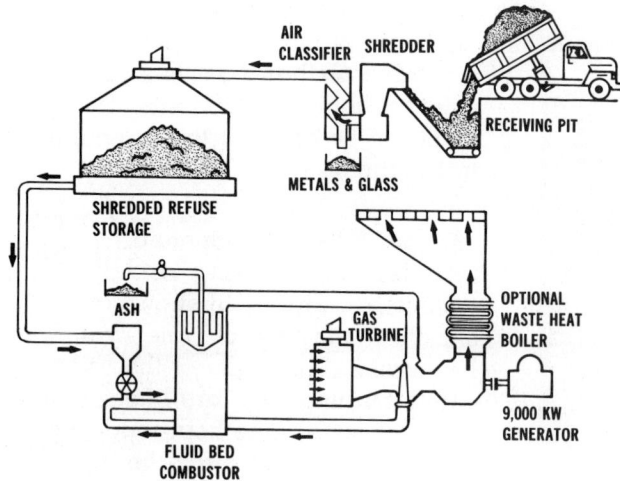

Fig. 34-14 Incinerator powered gas turbine.

waste powers a gas turbine generator to produce electricity.

11. Look at the drawing of the pyrolysis-type incineration system shown in Fig. 34-15. Pyrolysis chemically changes solid waste by heating it in an airless or oxygen-free chamber. The solid waste plastic produces ammonia, oxygen, hydrogen chloride, oil, tar, and many wax-like substances. The solid waste is first shredded. It is then air separated into metals and glass. The lighter waste passes into a second shredder. It then goes through the pyrolysis chamber. The byproducts of pyrolysis (solid char,

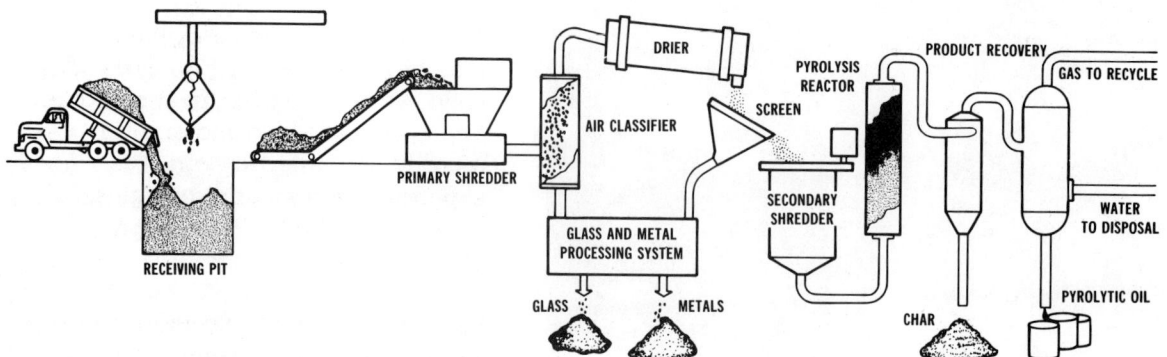

Fig. 34-15 Schematic of pyrolysis system.

waste water, organic liquids, and gases) are then collected. The gases can be burned for heating purposes.

12. Look at the drawing of a total incineration process shown in Fig. 34-16. In this process, the solid waste is incinerated at about 3000° F. The waste is melted into a slag and drained from the furnace. It is then solidified and some materials can be removed and recycled.

13. Look at the sketch of the fluid-bed incinerating process shown in Fig. 34-17. The solid waste is shredded, dried, and separated. The waste to be incinerated is mixed with air at 600° F. and is shot into the fluid-bed area. The exhaust gases are then removed from the two stages. The gas drives a turbine and generates electricity.

CONCLUSION

Incineration and sanitary landfill offer future ways of disposing of plastic solid wastes. Sanitary landfill has many advantages. For example, the completed landfill can often be used for park sites. There are generally no odors, underground fires, litter, or breeding of vermin in completed landfills. Landfilling is economical if the site is close to the community. Also, semi-skilled labor can be used to operate the landfill.

There are also some disadvantages to landfilling. Some communities must ship solid waste long distances to landfill sites. Landfill sites settle and produce methane gas for many years. This limits the type of construction to be placed on them. Water **must** be made to flow around landfills. If this is not done, underground water pollution will result.

Incineration also has many advantages. Incinerators can function in any type of weather.

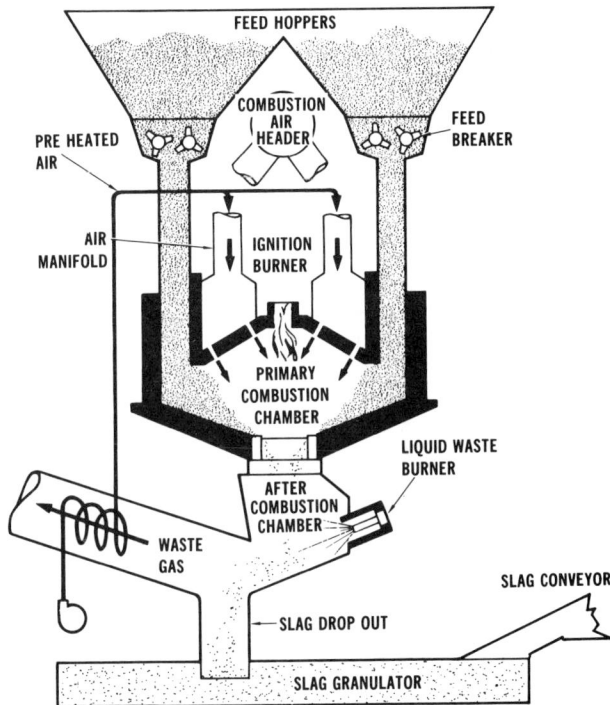

Fig. 34-16 Schematic of total incineration process.

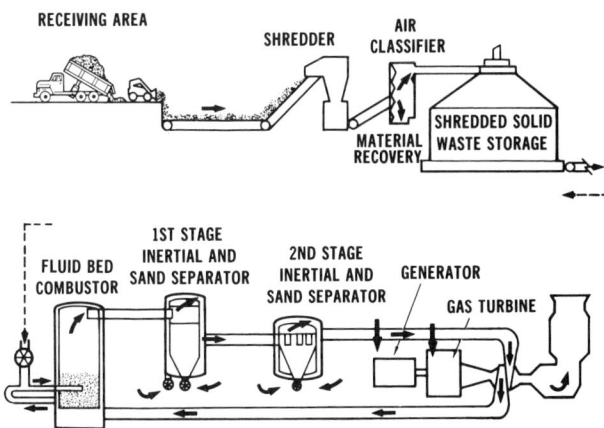

Fig. 34-17 Schematic of fluid-bed incinerator process.

Disposing

They can adjust to different daily amounts of solid waste. Only basic sorting of solid waste is needed before incineration. Less land and hauling distance is needed for incineration. The solid residue from incineration is small (about a 97% volume reduction) and free of germs. It also extends the sanitary landfill site life.

The byproducts of incineration can be sold to help pay for the cost of incineration. These products include steam, electricity, various gases, ashes, and wax-like items. With the cost of fuel increasing, incineration can help recover and lower energy costs. Because acceptable landfill sites are decreasing, incineration can reduce the amount of solid waste to be placed in landfills. This will extend the use of present landfill sites.

The disadvantages of incineration are as follows. Many incinerators air pollute. Wet waste must be burned using extra fuel. They are costly to build. The incineration waste must be sold or sanitary landfilled.

Some plastics pollute when incinerated or burned in open dumps. These plastics include PVC and polystyrene. Both produce smoke and PVC produces chlorine gas when poorly burned. Also, this chlorine mixes with water in the air to make hydrochloric acid.

Burning these plastics using modern incinerator technology will reduce their volume. It will also reduce air pollution.

When plastics are correctly burned, carbon dioxide and water are the main byproducts. Even hydrogen chloride gas can be controlled with proper incineration. When plastics are burned in open dumps and old incinerators, the pollution increases.

Section XI: Ecology

Often plastics can be incinerated at higher temperatures above the incinerator grate. This stops gumming up of the incinerator grates with hot plastic. Certain plastics have three to four times the heat content of other burning solid wastes. These plastics can be used as fuel to produce steam and electricity. Proper incineration technology along with pollution equipment will help reduce the solid wastes. They will also reduce air pollution.

REVIEW

1. Explain the differences among the area, trench, and ramp sanitary landfill methods.

2. Explain what happens to solid waste in the pyrolysis, water-wall, and fluid-bed incinerators.

3. List three advantages of sanitary landfilling solid waste.

4. List three advantages of incinerating solid waste.

5. Give a reason for compacting solid waste.

6. Give a reason for milling solid waste.

7. Explain why poorly incinerated PVC pollutes the air.

8. Explain why poorly incinerated polystyrene pollutes the air.

9. Describe how heat can be recovered from burning solid waste.

10. Name two uses for recovered incineration heat.

11. Explain why incineration resources must be recovered and used.

SELECTED BIBLIOGRAPHY

National Center for Resource Recovery, Inc. *Incineration.* Lexington, Massachusetts: Lexington Books, 1974.

————. *Municipal Solid Waste Collection.* Lexington, Massachusetts: Lexington Books, 1973.

————. *Sanitary Landfill.* Lexington, Massachusetts: Lexington Books, 1974.

Richardson, Terry A. *Modern Industrial Plastics.* Indianapolis: Howard W. Sams & Company, Inc., 1974.

Steele, Gerald L. *Exploring the World of Plastics.* Bloomington, Illinois: McKnight Publishing Company, 1977.

Articles:
"Air Classification," Fact Sheet Bulletin No. NCF-20-01. Washington, D.C.: National Center for Resource Recovery, Inc., November 1973.

"Let's Talk Trash," Bulletin No. GP-1973-2. Washington, D.C.: National Center for Resource Recovery, Inc.

"Municipal Solid Waste . . . A Source of Energy," Bulletin No. NCR-04-02. Washington, D.C.: National Center for Resource Recovery, Inc., Summer 1973.

"Plastics," Fact Sheet Bulletin No. NCF-14-01. Washington, D.C.: National Center for Resource Recovery, Inc., October, 1973.

"Plastics in Solid Waste," National Industrial Pollution Control Council: Sub-Council Report, March, 1971, Washington, D.C.

Glossary

Additives Substances blended with plastics to improve various properties. An example is a stabilizer added to prevent plastic from degrading when exposed to heat and sunlight.

Air pollution Filling the atmosphere with undesirable toxic and nontoxic particles.

A-stage An early stage in the reaction of a thermosetting resin. At this stage, the material is still soluble in certain liquids and fusible.

ASTM American Society for Testing Materials.

Base material Plastic sheets or strips being welded. Also called parent material.

Biaxial Having two directions or two axes.

Bisect To divide into two parts or in half.

Blow molding A method of fabrication in which a parison (hollow tube) is forced into the shape of a mold cavity by internal air pressure.

Bodied cement See DOPE CEMENT.

B-stage An intermediate stage in the reaction of a thermosetting resin. At this stage, the material softens when heated and swells in contact with certain liquids, but does not entirely fuse or dissolve. Resins in thermosetting molding compounds are usually in this stage.

Bulking To increase the mass of an object.

Bulking agent Material mixed with resin to increase its density and volume, thereby reducing the cost of the finished mold.

Buoyant The ability of an article or material to float in a liquid.

Burn test The application of heat (flame or electrical) to a plastic to determine its type.

Capillary cementing A chemical welding process in which the solvent cement is drawn between the plastic pieces and into the joint.

Cast To form a plastic object by pouring a liquid monomer-polymer solution into an open mold where it finishes polymerization.

Casting The finished product of a casting operation. Not synonymous with the word "molding."

Catalyst A substance which accelerates a reaction in a substance. For example, adding a catalyst will increase the curing speed of a compound.

Cellular Resins in sponge form. Also called expanded plastic or foamed plastic. The sponge may be flexible or rigid, the cells closed or interconnected, and the density ranging from that of the solid parent resin down to 2 lbs. per cubic foot.

Cellulosics A family of plastics with the main constituent being the polymeric carbohydrate cellulose.

Chain A long, heavy linear molecule.

Charge The measurement of weight of material used to load a mold at one time or during one cycle.

Chase A device that holds the die or type in the hot foil stamping machine.

Chemical properties The characteristics of a material that enable it to withstand chemical forces. An example is corrosion resistance.

Chiller The chilling system for the extrudate in the extrusion process. See EXTRUSION.

Clear casting Pouring a catalyzed liquid plastic into an open mold.

Closed-celled Cells not interconnected. The condition of cells which make up cellular or foamed plastics.

Cold dip coating A coating process in which a tool or part at room temperature is coated with a peelable plastic. The item is dipped into the heated plastic. Also called cold dipping.

Colorant A dye or pigment used to impart color to plastics.

Composite A material compound composed of distinct parts with each having distinct mechanical, physical, and chemical properties.

Composting Grinding trash into small pieces and mixing it with soil.

Concave An inwardly curved surface.

Crosslinking As applied to polymer molecules, it is the creating of chemical links between the molecular chains. When extensive, as in most thermosetting resins, crosslinking makes one infusible supermolecule of all chains.

C-stage The final stage in the reactions of a thermosetting resin. At this stage, the material is relatively insoluble and infusible. Thermosetting resins in a fully cured plastic are in this stage.

Cure To change the physical properties of a material by chemical reaction. Curing is usually accomplished by the action of heat and catalysts, alone or in combination, with or without pressure.

Curing time In the molding of thermosetting plastics, the interval of time between the instant the relative movement between the moving parts of a mold stop and the instant that pressure is released.

Cycle The completed sequence of events in a process or operation.

Cycle time In molding, the cycle time is the period or elapsed time between a certain point in one cycle and the same point in the next cycle.

Deflashing Removing the flash on a plastic molding.

Degassing Removing trapped air from a material. Breathing or the brief opening and closing of the platens to allow moisture vapor to escape during the laminating process.

Dehydrate To dry.

Delamination The separation of layers of laminate caused by the failure of the adhesive.

Demold To eject or remove a part or pattern from a mold.

Density Weight per unit volume of a substance expressed in grams per cubic centimeter or pounds per cubic foot.

Die (1) The orifice through which hot plastic is forced in the extrusion process. The die determines the shape of the extruded profile or product; (2) a decorative design clamped in the hot foil stamping machine chase; (3) the sealing bar

area of the impulse sealing machine; (4) the pattern or grooved type fastened to the machine engraver work table.

Die head A device to which the blow mold is attached. It aids in changing the plastic flow from a horizontal to a vertical downward movement to the die.

Dip chemical welding A chemical welding process in which the edges of the plastic pieces to be joined are first dipped into solvent cement.

Dip coating Applying a plastics coating by dipping the object to be coated into a tank of melted resin or plastisol.

Dispersion Finely divided particles of a material in suspension in another material.

Dope cement A cement composed of solvents and a small quantity of the plastic to be joined. Sometimes called bodied cement.

Draft The degree of taper of a side wall or the angle of clearance designed to facilitate removal of parts from a mold.

Dry printing Another name for hot foil stamping. See HOT FOIL STAMPING.

Dwell The amount of time the die (under pressure) contacts the foil or leaf material in hot foil stamping.

Eject To remove a part from the mold cavity.

Elastomer A material which stretches under low stress at room temperature to at least twice its length and snaps back to its original length upon release of stress.

Elongation Stretching or lengthening.

Embed To place in a mass of material.

Emulsion A suspension of one fluid in another.

Encapsulating Pouring a catalyzed liquid plastic around a sample in a mold.

Engraving Cutting figures, letters, or designs into a surface.

Epoxy resins Based on ethylene oxide, its derivatives, or homologs, epoxy resins form straight-chain thermoplastics and thermosetting resins.

Exotherm The amount of heat given off. The temperature time curve of a chemical reaction giving off heat, particularly the polymerization of casting resins.

Extrudate The product or material delivered by an extruder.

Extrusion The compacting of a plastics material and the forcing of it through an orifice in more or less continuous fashion.

Extrusion blow molding A blow molding process in which the parison to be blown is formed by extrusion molding.

Ferrous metals Metals containing some iron.

Filler A cheap, inert substance added to a plastic to make it less costly. Fillers may also improve physical properties, particularly hardness, stiffness, and impact strength.

Filler rod Profile shapes made of the base material and used to fill the joint between the base materials.

Fillet A narrow length of concave surface.

Film Sheeting having a nominal thickness not greater than one ten-thousandths of an inch.

Free forming Using air pressure to blow a heated sheet of plastic being held in a frame until the desired shape or height is attained.

Fuse To heat materials until they become one homogeneous substance.

Fusion The act of melting or fusing together.

Gate In injection molding, the opening through which the melted plastic enters the mold cavity.

Gel coat A thin layer of resin, sometimes containing pigment, applied to a reinforced plastics molding as a cosmetic.

Gel time The time needed for a cast liquid to turn into a jelly-like substance.

Gradient density column A means of measuring the densities of small plastics samples.

Hand tool heat sealing A technique of bonding thermoplastics using a hand operated clamp, wheel, or blade to apply heat and pressure to plastic areas in contact.

Hand welding Hot air welding with a rod in one hand and a torch in the other.

Hardener An additive that promotes or controls the curing reaction. The term is also used to refer to an additive used to control the degree of hardness of the cured film. Often used synonymously with the term catalyst. See CATALYST.

Fixture A device attached to the hot foil stamping machine base. It holds the product being stamped in position.

Flash Excess plastic attached to a molding along the parting line. It must be removed before the part can be considered finished.

Flash mold A compression mold designed so that excess compression molding compound (flash) can be forced out of the molding cavity as it is closed.

Flexible foam A pliable foam with an open-cell internal structure.

Flexible molds Molds made of rubber or elastomeric plastics. They are used for casting plastics and can be stretched to remove cured pieces with undercuts.

Flexural Test A test that measures the ability of a plastic to bend or flex.

Fluidized bed coating A method of applying a coating of thermoplastic resin to an article in which the article is immersed in a fluidized bed of powdered resin and then heated in an oven.

Foaming The process of producing a cellular or foamed product.

Foaming agents Chemicals added to plastics that generate inert gases on heating, causing the resin to assume a cellular structure.

Foil Hot stamping decorative film or leaf. Composed of many different layers of material.

Hardening Synonymous with the term cure. See CURE.

Hardening time Period of time starting at the end of the pot life period and ending when the casting is rigid.

Heating element Electrical air, or gas heater located in the hot air welding gun barrel.

Heat sealing A method of joining plastics by the simultaneous application of heat and pressure to areas in contact. This method is also called heat joining.

Heat shrink tubing Plastic tubing designed to shrink when heated.

High pressure laminates Laminates molded and cured at pressures not lower than 1000 psi and more commonly in the range of 1200 to 2000 psi.

Hot dip coating A coating process in which a heated mold or product is dipped into and coated with a plastic dispersion. This process is also called hot dipping.

Hot gas welding A technique of joining thermoplastics materials (usually sheet) by directing a jet of hot air from a welding torch against the plastic area to be joined.

Hot leaf stamping A decorating operation for marking plastics in which a roll leaf is stamped with heated metal dies onto the face of the plastics.

Hot wire "L" sealer A machine with two sealing units at right angles to each other. This allows two sides of a plastic package to be hot wire sealed at the same time.

Hot wire sealing A heat sealing technique in which a seal is formed by applying heat and pressure at the same time.

Hydroponics Growing plants by placing their roots in liquid nutrients.

Impact strength The ability of a material to withstand the application of sudden force (shock loading).

Impregnate To thoroughly soak a material with synthetic resin so that the resin gets within the body of the material.

Impregnation The process of impregnating a material.

Impulse sealing A heat sealing technique in which a pulse of intense thermal energy is applied to the sealing area for a very short time, followed immediately by cooling.

Incineration The burning of waste materials in a specially designed incinerator.

Incising Cutting or scribing lines into a surface.

Inert Inactive or possessing no inherent power or action.

Injection blow molding A blow molding process in which the parison to be blown is formed by injection molding.

Injection mold A mold into which a plasticated material is introduced from the exterior heating cylinder.

Injection molding A molding process in which a heat-softened plastic is forced into a relatively cool cavity to give it the desired shape.

Laminated plastic A plastic material consisting of superimposed layers of synthetic resin-impregnated or coated filler. The layers have been bonded together by means of heat and pressure to form a single piece.

Laminating cement See DOPE CEMENT.

Lap joint A joint formed by laying one plastic edge over another and then bonding the two together.

Latex An emulsion of natural or synthetic resin particles dispersed in a watery medium.

Low pressure laminates Laminates formed and cured with pressures less than 1000 psi.

Mandrels Molds used for hot dip coating. Also called plugs.

Matched mold Close fitting molds between which are molded reinforced plastic products.

Mechanical forming Shaping or forming heated sheets of plastic by hand or with the aid of jigs and fixtures. No mold is used.

Mechanical properties Characteristics which enable a material to withstand mechanical forces. A few examples are strength, hardness, impact, and ductility.

Modeling clay Nonhardening clay substance.

Mold To shape plastic parts for finished products by heat and pressure. A cavity into which melted plastics are poured to form a product.

Mold breathing See DEGASSING.

Molding compound See CHARGE.

Molding cycle The complete period of time required for the production of one set of moldings.

Molding powder Plastics material in varying stages of granulation ready for use in the molding operation.

Molding pressure The pressure used in an injection machine or press to force the softened plastic into the mold cavities.

Molecule The smallest particle of a substance that can exist independently and still retain the chemical properties of the substance.

Mulching Synonymous with or similar to composting. See COMPOSTING.

Negative draft Synonymous with undercut. See UNDERCUT.

Open—celled Interconnected cells. The interconnecting of cells in cellular or foamed plastics.

Open dumping Placing trash or waste in open, uncontrolled areas on the land.

Organic Chemical compounds derived from living organisms or compounds of carbon.

Organosol A vinyl or nylon dispersion. The liquid phase contains one or more organic solvents.

Orientated film Plastic that is stretched in a specific direction during its initial processing.

Pantograph A machine used to mechanically duplicate designs.

Parallelogram A quadrilateral shape in which opposite sides are parallel.

Parent material See BASE MATERIAL.

Parison The tube shape of the plastic formed by the die in blow molding.

Parting line Mark on a molding or casting where mold halves have met in closing.

Pattern A model over and around which material is cast.

Pelletizer A machine that chops the emerging extrudate into pellets.

Physical foaming Mixing gas with plastic through physical means and causing the gas to increase its volume within a heated and softened plastic. This action produces a cellular plastic such as those produced with polystyrene expandable beads.

Physical properties Characteristics which enable a material to withstand physical forces. A few examples are heat, conductivity, light reflectivity, and electrical conductivity.

Pinch-off A raised edge around the cavity in the mold. It seals off the part and separates the excess material as the mold closes around the parison.

Pinhole A very small hole in an extruded resin coating.

Plasticate To soften by heating or kneading.

Plastic flow The fusing and spreading of plastic particles over a mold surface.

Plasticize To soften a material and make it plastic or moldable either by means of a plasticizer or by the application of more heat.

Plasticizer A chemical agent added to plastics compositions to make them softer and more flexible.

Plastisols Mixtures of resins and plasticizers which can be molded, cast, or converted to continuous films by the application of heat. If the mixtures contain volatile thinners, they are also known as organosols. See ORGANOSOLS.

Platens The mounting plates of a press to which the entire mold assembly is bolted.

Plugs See MANDRELS.

Pock-marked Discolored surface spots or areas.

Pollute To contaminate or make unclean.

Polyester A resin formed by the reaction between a dibasic acid and a dihydroxy alcohol, both organic.

Polymerization The process of growing large molecules from small ones.

Polyurethane A family of resins produced by reacting di-isocyanate with organic compounds containing two or more active hydrogens to form polymers having free isocyanate groups.

Positive mold A mold designed to trap all the molding materials when it closes.

Post-weld stress cracking The cracking of certain thermoplastics after they have been welded.

Pot life The length of time required to mix silicone components before they start to harden (silicone molds).

Potting Similar to encapsulating except that steps are taken to ensure complete penetration of all the voids in the object before the resin polymerizes.

Pre-expanded beads Gas-containing beads which have been heated, softened, and the internal gas volume increased to produce a larger than original size bead.

Preform To make plastic molding powders into pellets or tablets. A compressed tablet.

Profile The cross-sectional shape of an extrusion product.

Puller An extrudate pulling machine.

Purging Cleaning a color or type of material from the cylinder of an injection molding machine by forcing it out with the new color or material to be used in the next operation. Purging materials are also available.

PVA Polyvinyl alcohol. A water soluble release agent containing a coloring agent.

PVC Polyvinyl chloride. A thermoplastic material composed of polymers of vinyl chloride.

Pyrolysis A process used to change waste chemically into usable compounds.

Radio frequency welding A method of welding thermoplastics using a radio frequency field to apply the necessary heat. Also called high frequency welding.

Reciprocating screw An extruding system in which the movement of a rotating screw collects the molten polymer and forces it through the head and die at high speed.

Recycle To reuse or reprocess.

Reinforcement A strong inert material bound into a plastic to improve its strength, stiffness, and impact resistance.

Release agent A lubricant (usually wax) used to coat a mold cavity to prevent the product from sticking to the mold.

Resin Any of a class of solid or semisolid organic products of natural or synthetic origin. Resins are generally of high molecular weight with no definite melting point. Most resins are polymers.

Retainer A flask, chase, or material container.

Rigid foam A stiff foam with a closed-celled internal structure.

Root Inside bottom of an unwelded joint.

Root gap Distance between the edges of the inside bottom of an unwelded joint.

Rotational Molding A plastic molding process in which two molds filled with heated plastic powders or liquids are rotated at the same time in two planes at right angles to one another.

Roving Strands of reinforcement material wound together on a spool.

RTV Room temperature vulcanizing.

Runner Network of channels placed in a mold to allow casting material to reach the mold cavity.

Sandwich A laminated construction consisting of thin facings bonded to a relatively thick lightweight core.

Sanitary landfill The controlled filling of lowlands or trenches with solid waste.

Scar A characteristic mark usually found on the bottom of plastic containers. It is caused by the pinch-off operation and is often referred to as the "length of the pinch-off." See PINCH-OFF.

Sealer coat A finishing coat placed on the pattern, molding board, and retainer box to prevent them from degassing and chemically reacting with the resin.

Self-extinguishing The ability of some plastics to self-eliminate a flame spread. See THERMOSET.

Semipositive mold A mold which allows a small amount of excessive materials to escape when it is closed.

Sheet plastic A flat section of thermoplastic resin approximately 10 mils or greater in thickness.

Shrink wrapping A packaging technique in which the strains in a plastic film are released by raising the temperature of the film, thereby causing it to shrink over the package.

Silicone rubber A polymeric material in which the recurring chemical group contains silicone and oxygen atoms as links in the main chain.

Sizer plate The split plate mounted on the water tank in an extruder. It is used to control the size of the extrudate.

Solid waste Concrete, bricks, many plastics, and other materials that do not rot or decay.

Solubility Capable of being dissolved.

Solubility Test Testing the solubility of a plastic. Sometimes called the Solvent Test.

Solvent A material or substance that has the capability of dissolving other materials or substances.

Solvent cement A cement which dissolves the plastic being joined and forms strong intermolecular bonds as it evaporates.

Solvent mix See DOPE CEMENT.

Spaghetti tubing Same as heat shrink tubing, but smaller in diameter.

Specific gravity The ratio of the weight of a volume of a substance (here a plastic) to the weight of an equal volume of water at 73° F.

Split pattern A symmetrical two-part

Glossary

model. One part or half is a mirror image of the other half.

Sprayup A plastics fabrication process in which a spray gun is used to spray resin and a reinforcement material into a mold cavity.

Sprue The pouring channel placed in a mold.

Stress The internal resistance of a part or material to the external forces applied to it.

Stylus A pointed instrument or guide that follows die grooves or contours.

Suspension The fine particles of any solid mixed but undissolved in a liquid, or gas.

Tacker An electric hand-held sealing iron. It is used to heat seal plastic packages by hand.

Templet A pattern or guide used to provide shape to the product or product areas being formed in a mold.

Tensile strength The ability of a material to resist being pulled apart.

Tensile test A test of the tensile strength of a material.

Thermoelasticity The physical property of becoming elastic when heated and rigid when cooled.

Thermoforming Any process of forming thermoplastic sheet which consists of heating the sheet and pulling it down onto a mold surface.

Thermoplastic Any plastic that melts under the application of heat.

Thermoset Any plastic that resists burning or melting under the application of heat.

Tracer fiber A nonfilament of colored fiber glass composing part of the roving. Aids the chopper operator in determining the uniformity of chopped fiber over a mold surface.

Tumbling Removing gates, flash, and fins from small plastic objects by rotating them in a barrel or on a belt together with wooden pegs, sawdust, and some polishing compounds.

Undercuts A protuberance or indentation that impedes withdrawal from a two-piece rigid mold.

Vacuum forming A method of sheet forming in which the plastic is clamped in a stationary frame, heated, and drawn down by a vacuum into a mold.

Viscos Extremely thick or jelly-like.

Vortex A whirlpool of material forming a vacuum at its center.

Water pollution Waste products placed into rivers and waterways.

Ways Special workable edges machined to accept type holder slides.

Weld To unite, join, or fuse into one piece.

Welding Joining thermoplastics by one of several heat-softening or chemical-softening processes.

Wetout Working resin through and air out of reinforcing material.

Whirler A hand-operated rotating bench wheel on which a mold is placed. Aids the operator in obtaining a uniform mold deposit of chopped fibers and sprayed resin.